Emergency Management

Emergency Management

Edited by
Conner Long

Larsen & Keller
www.larsen-keller.com

Emergency Management
Edited by Conner Long
ISBN: 978-1-63549-102-9 (Hardback)

Larsen & Keller
Published by Larsen and Keller Education,
5 Penn Plaza,
19th Floor,
New York, NY 10001, USA

Cataloging-in-Publication Data

Emergency management / edited by Conner Long.
p. cm.
Includes bibliographical references and index.
ISBN 978-1-63549-102-9
1. Emergency management. 2. Disaster relief. 3. Environmental disasters.
I. Long, Conner.
HV551.2 .E44 2017
363.348--dc23

Printed and bound in China.

For more information regarding Larsen and Keller Education and its products, please visit the publisher's website www.larsen-keller.com

Table of Contents

Preface

Emergency management or disaster management is a field which deals with the management of situations when a natural or man made disaster strikes. It includes making plans and schemes to minimize the impact of the hazard and control the rate of mortality and other losses. This book aims at providing the students with thorough insights into the field of emergency management. It explains the complex subject in the most comprehensible and easy to understand language. Most of the topics introduced in this textbook cover new techniques, technologies and the applications of emergency management. Coherent flow of topics, student-friendly language and extensive use of examples make this text an invaluable source of knowledge.

A detailed account of the significant topics covered in this book is provided below:

Chapter 1- Emergency is any situation that causes risk to life, property or the environment. It requires intervention in order to control the situation. Some emergencies are caused naturally while some are caused by humans. This chapter is an overview of the subject matter incorporating all the major aspects of emergency and emergency management.

Chapter 2- Natural disasters are disasters that are caused naturally. Some of the examples of natural disasters are earthquakes, sinkholes, limnic eruptions and tsunamis. This section also discusses meteorological disasters such as blizzards, cyclones, droughts and thunderstorms. The topics explained are of great importance to broaden the existing knowledge on natural disasters.

Chapter 3- Man-made disasters are disasters which are the consequences of technological hazards. The examples of man-made disasters are terrorism, sabotage, ecotage, genocides, riots and massacres. Wars and deliberate attacks are also a part of this category of disasters. This section discusses man-made disasters in a critical manner providing key analysis to the subject matter.

Chapter 4- A particular type of bunker that is used to protect the occupants from severe weather conditions is known as a storm shelter. Storm shelters are usually seen in the Midwest and southeast United States. The alternative strategies used in emergency mitigation are earthquake valve, fallout shelter, disaster recovery plan, dam safety system, community resilience etc. The topics elucidated in this section are of vital importance, and provides a better understanding of emergency mitigation strategies.

Chapter 5- To develop a global understanding of emergency, it is important to mention agencies such as United States Environmental Protection Agency, Local Emergency Planning Committee, the World Conference on Disaster Risk Reduction and Federal Disaster Assistance Nonprofit Fairness Act of 2013. This chapter is an overview of the subject matter incorporating all the major aspects of emergency management.

I would like to make a special mention of my publisher who considered me worthy of this opportunity and also supported me throughout the process. I would also like to thank the editing team at the back-end who extended their help whenever required.

Editor

1

Introduction to Emergency and Emergency Management

Emergency is any situation that causes risk to life, property or the environment. It requires intervention in order to control the situation. Some emergencies are caused naturally while some are caused by humans. This chapter is an overview of the subject matter incorporating all the major aspects of emergency and emergency management.

Emergency

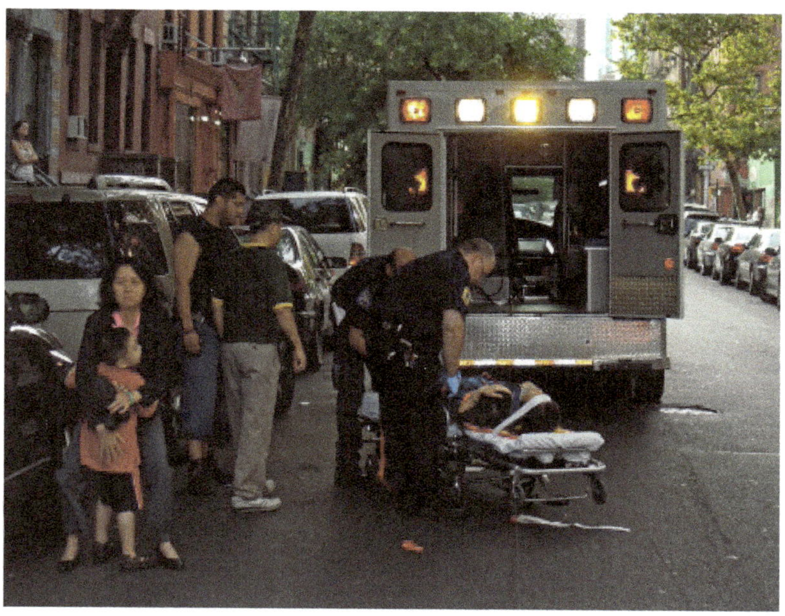

Paramedics treating a woman who has collapsed in the street in New York. Dangers to life and health are serious enough that emergency response systems are considered vital.

An emergency is a situation that poses an immediate risk to health, life, property, or environment. Most emergencies require urgent intervention to prevent a worsening of the situation, although in some situations, mitigation may not be possible and agencies may only be able to offer palliative care for the aftermath.

While some emergencies are self-evident (such as a natural disaster that threatens many lives), many smaller incidents require that an observer (or affected party) decide whether it qualifies as an emergency. The precise definition of an emergency, the agencies involved and the procedures used, vary by jurisdiction, and this is usually set by the government, whose agencies (emergency services) are responsible for emergency planning and management.

Defining an Emergency

An incident, to be an emergency, conforms to one or more of the following: if it:

- Poses an immediate threat to life, health, property, or environment
- Has already caused loss of life, health detriments, property damage, or environmental damage
- has a high probability of escalating to cause immediate danger to life, health, property, or environment

In the United States, most states mandate that a notice be printed in each telephone book that requires that someone must relinquish use of a phone line, if a person requests the use of a telephone line (such as a party line) to report an emergency. State statures typically define an *emergency* as, "...a condition where life, health, or property is in jeopardy, and the prompt summoning of aid is essential."

Whilst most emergency services agree on protecting human health, life and property, the environmental impacts are not considered sufficiently important by some agencies. This also extends to areas such as animal welfare, where some emergency organisations cover this element through the "property" definition, where animals owned by a person are threatened (although this does not cover wild animals). This means that some agencies do not mount an "emergency" response where it endangers wild animals or environment, though others respond to such incidents (such as oil spills at sea that threaten marine life). The attitude of the agencies involved is likely to reflect the predominant opinion of the government of the area.

Types of Emergency

Dangers to Life

Many emergencies cause an immediate danger to the life of people involved. This can range from emergencies affecting a single person, such as the entire range of medical emergencies including heart attacks, strokes, cardiac arrest and trauma, to incidents that affect large numbers of people such as natural disasters including tornadoes, hurricanes, floods, earthquakes, mudslides and outbreaks of diseases such as cholera, Ebola, and malaria.

Most agencies consider these the highest priority emergency, which follows the general school of thought that nothing is more important than human life.

Dangers to Health

Some emergencies are not necessarily immediately threatening to life, but might have serious implications for the continued health and well-being of a person or persons (though a health emergency can subsequently escalate to life-threatening).

The causes of a health emergency are often very similar to the causes of an emergency threatening to life, which includes medical emergencies and natural disasters, although the range of incidents that can be categorised here is far greater than those that cause a danger to life (such as broken limbs, which do not usually cause death, but immediate intervention is required if the person is to recover properly). Many life emergencies, such as cardiac arrest, are also health emergencies.

Dangers to the Environment

Some emergencies do not immediately endanger life, health or property, but do affect the natural environment and creatures living within it. Not all agencies consider this a genuine emergency, but it can have far reaching effects on animals and the long term condition of the land. Examples would include forest fires and marine oil spills.

Systems of Classifying Emergencies

Agencies across the world have different systems for classifying incidents, but all of them serve to help them allocate finite resource, by prioritising between different emergencies.

The first stage of any classification is likely to define whether the incident qualifies as an emergency, and consequently if it warrants an emergency response. Some agencies may still respond to non-emergency calls, depending on their remit and availability of resource. An example of this would be a fire department responding to help retrieve a cat from a tree, where no life, health or property is immediately at risk.

Following this, many agencies assign a sub-classification to the emergency, prioritising incidents that have the most potential for risk to life, health or property (in that order). For instance, many ambulance services use a system called the Advanced Medical Priority Dispatch System (AMPDS) or a similar solution. The AMPDS categorises all calls to the ambulance service using it as either 'A' category (immediately life-threatening), 'B' Category (immediately health threatening) or 'C' category (non-emergency call that still requires a response). Some services have a fourth category, where they believe that no response is required after clinical questions are asked.

Another system for prioritizing medical calls is known as Emergency Medical Dispatch (EMD). Jurisdictions that use EMD typically assign a code of "alpha" (low priority), "bravo" (medium priority), "charlie" (requiring advanced life support), delta (high priority, requiring advanced life support) or "echo" (maximum possible priority, e.g., witnessed cardiac arrests) to each inbound request for service; these codes are then used to determine the appropriate level of response.

Other systems (especially as regards major incidents) use objective measures to direct resource. Two such systems are SAD CHALET and ETHANE, which are both mnemonics to help emergency services staff classify incidents, and direct resource. Each of these acronyms helps ascertain the number of casualties (usually including the number of dead and number of non-injured people involved), how the incident has occurred, and what emergency services are required.

Agencies Involved in Dealing with Emergencies

Most developed countries have a number of emergency services operating within them, whose purpose is to provide assistance in dealing with any emergency. They are often government operated, paid for from tax revenue as a public service, but in some cases, they may be private companies, responding to emergencies in return for payment, or they may be voluntary organisations, providing the assistance from funds raised from donations.

Most developed countries operate three core emergency services:

- Police – who deal with security of person and property, which can cover all three categories of emergency. They also to some extent deal with punishment of those who cause an emergency through their deliberate actions.
- Fire service – who deal with potentially harmful fires, but also often rescue operations such as dealing with road traffic collisions. Their actions help to prevent loss of life, damage to health and damage to or loss of property.
- Emergency Medical Service (Ambulance / Paramedic service) – These services attempt to reduce loss of life or damage to health. This service is likely to be decisive in attempts to prevent loss of life and damage to health. In some areas "Emergency Medical Service" is abbreviated to simply EMS.

In some countries or regions, two or more of these services may be provided by the same agency (e.g. the fire service providing emergency medical cover), and under different conditions (e.g. publicly funded fire service and police, but a private ambulance service)

There may also be a number of secondary emergency services, which may be a part of one of the core agencies, or may be separate entities who assist the main agencies. This can include services providing specialist rescue (such as mountain rescue or mine rescue), bomb disposal or search and rescue.

The Military and the Amateur Radio Emergency Service (ARES) or Radio Amateur Civil Emergency Service (RACES) help in large emergencies such as a disaster or major civil unrest.

Summoning Emergency Services

Most countries have an emergency telephone number, also known as the universal emergency number, which can be used to summon the emergency services to any incident. This number varies from country to country (and in some cases by region within a country), but in most cases, they are in a short number format, such as 911 (United States and many parts of Canada), 999 (United Kingdom), 112 (Europe) and 000 (Australia).

The majority of mobile phones also dial the emergency services, even if the phone keyboard is locked, or if the phone has an expired or missing SIM card, although the provision of this service varies by country and network.

Civil Emergency Services

In addition to those services provided specifically for emergencies, there may be a number of agencies who provide an emergency service as an incidental part of their normal 'day job' provision. This can include public utility workers, such as in provision of electricity or gas, who may be required to respond quickly, as both utilities have a large potential to cause danger to life, health and property if there is an infrastructure failure

Domestic Emergency Services

Generally perceived as pay per use emergency services, domestic emergency services are small, medium or large businesses who tend to emergencies within the boundaries of licensing or capabilities. These tend to consist of emergencies where health or property is perceived to be at risk but may not qualify for official emergency response. Domestic emergency services are in principal similar to civil emergency services where public or private utility workers will perform corrective repairs to essential services and avail their service at all times however these are at a cost for the service. An example would be an emergency plumber

Emergency Action Principles (EAP)

Emergency action principles are key 'rules' that guide the actions of rescuers and potential rescuers. Because of the inherent nature of emergencies, no two are likely to be the same, so emergency action principles help to guide rescuers at incidents, by sticking to some basic tenets.

The adherence to (and contents of) the principles by would-be rescuers varies widely based on the training the people involved in emergency have received, the support available from emergency services (and the time it takes to arrive) and the emergency itself.

Key Emergency Principle

The key principle taught in almost all systems is that the rescuer, be they a lay person or a professional, should assess the situation for danger.

The reason that an assessment for danger is given such high priority is that it is core to emergency management that rescuers do not become secondary victims of any incident, as this creates a further emergency that must be dealt with.

A typical assessment for danger would involve observation of the surroundings, starting with the cause of the accident (e.g. a falling object) and expanding outwards to include any situational hazards (e.g. fast moving traffic) and history or secondary information given by witnesses, bystanders or the emergency services (e.g. an attacker still waiting nearby).

Once a primary danger assessment has been complete, this should not end the system of checking for danger, but should inform all other parts of the process.

If at any time the risk from any hazard poses a significant danger (as a factor of likelihood and seriousness) to the rescuer, they should consider whether they should approach the scene (or leave the scene if appropriate).

Managing an Emergency

There are many emergency services protocols that apply in an emergency, which usually start with planning before an emergency occurs. One commonly used system for demonstrating the phases is shown here on the right.

The planning phase starts at preparedness, where the agencies decide how to respond to a given incident or set of circumstances. This should ideally include lines of command and control, and division of activities between agencies. This avoids potentially negative situations such as three separate agencies all starting an official rest centre for victims of a disaster.

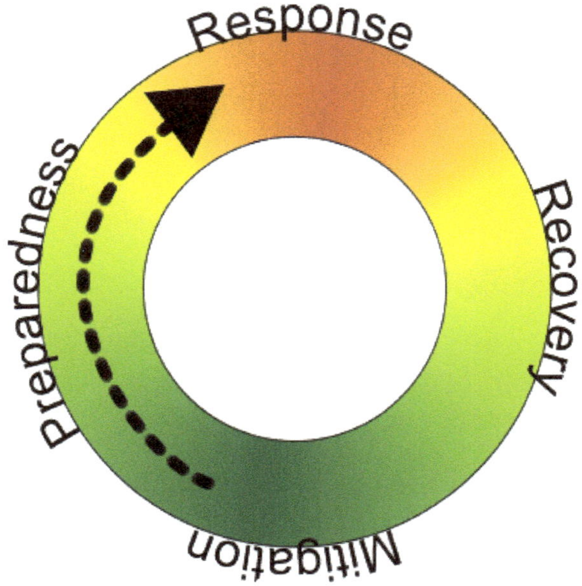

A graphic representation of the four phases in emergency management.

Following an emergency occurring, the agencies then move to a response phase, where they execute their plans, and may end up improvising some areas of their response (due to gaps in the planning phase, which are inevitable due to the individual nature of most incidents).

Agencies may then be involved in recovery following the incident, where they assist in the clear up from the incident, or help the people involved overcome their mental trauma.

The final phase in the circle is mitigation, which involves taking steps to ensure no re-occurrence is possible, or putting additional plans in place to ensure less damage is done. This should feed back in to the preparedness stage, with updated plans in place to deal with future emergencies, thus completing the circle.

State of Emergency

In the event of a major incident, such as civil unrest or a major disaster, many governments maintain the right to declare a state of emergency, which gives them extensive powers over the daily lives of their citizens, and may include temporary curtailment on certain civil rights, including the right to trial (for instance to discourage looting of an evacuated area, a shoot on sight policy, however unlikely, may be in force)

Disaster

A disaster is a serious disruption of the functioning of a community or a society involving widespread human, material, economic or environmental losses and impacts, which exceeds the ability of the affected community or society to cope using its own resources.

In contemporary academia, disasters are seen as the consequence of inappropriately managed risk. These risks are the product of a combination of both hazards and vulnerability. Hazards that strike in areas with low vulnerability will never become disasters, as in the case of uninhabited regions.

Ruins from the 1906 San Francisco earthquake, remembered as one of the worst natural disasters in the history of the United States.

Developing countries suffer the greatest costs when a disaster hits – more than 95 percent of all deaths caused by hazards occur in developing countries, and losses due to natural hazards are 20 times greater (as a percentage of GDP) in developing countries than in industrialized countries.

Classifications

Researchers have been studying disasters for more than a century, and for more than forty years disaster research. The studies reflect a common opinion when they argue that all disasters can be seen as being human-made, their reasoning being that human actions before the strike of the hazard can prevent it developing into a disaster. All disasters are hence the result of human failure to introduce appropriate disaster management measures. Hazards are routinely divided into natural or human-made, although complex disasters, where there is no single root cause, are more common in developing countries. A specific disaster may spawn a secondary disaster that increases the impact. A classic example is an earthquake that causes a tsunami, resulting in coastal flooding.

Natural Hazard

A natural hazard is a natural process or phenomenon that may cause loss of life, injury or other health impacts, property damage, loss of livelihoods and services, social and economic disruption, or environmental damage.

Various phenomena like earthquakes, landslides, volcanic eruptions, floods, hurricanes, tornadoes, blizzards, tsunamis, and cyclones are all natural hazards that kill thousands of people and destroy billions of dollars of habitat and property each year. However, the rapid growth of the world's population and its increased concentration often in hazardous environments has escalated both the frequency and severity of disasters. With the tropical climate and unstable land forms, coupled with deforestation, unplanned growth proliferation, non-engineered constructions which make the disaster-prone areas more vulnerable, tardy communication, and poor or no budgetary allocation for disaster prevention, developing countries suffer more or less chronically from natural disasters. Asia tops the list of casualties caused by natural hazards.

Airplane crashes and terrorist attacks are examples of man-made disasters: they cause pollution, kill people, and damage property. This example is the September 11 attacks in 2001 at the World Trade Center in New York.

Human-instigated

Human-instigated disasters are the consequence of technological hazards. Examples include stampedes, fires, transport accidents, industrial accidents, oil spills and nuclear explosions/radiation. War and deliberate attacks may also be put in this category. As with natural hazards, man-made hazards are events that have not happened—for instance, terrorism. Man-made disasters are examples of specific cases where man-made hazards have become reality in an event.

Emergency Management

Disaster management (or emergency management) is the creation of plans through which communities reduce vulnerability to hazards and cope with disasters. Disaster management does not avert or eliminate the threats; instead, it focuses on creating plans to decrease the effect of disasters. Failure to create a plan could lead to human mortality, lost revenue, and damage to assets. Currently in the United States 60 percent of businesses do not have emergency management plans. Events covered by disaster management include acts of terrorism, industrial sabotage, fire, natural disasters (such as earthquakes, hurricanes, etc.), public disorder, industrial accidents, and communication failures.

Emergency Planning Ideals

If possible, emergency planning should aim to prevent emergencies from occurring, and failing that, should develop a good action plan to mitigate the results and effects of any emergencies. As time goes on, and more data becomes available, usually through the study of emergencies as they occur, a plan should evolve. The development of emergency plans is a cyclical process, common to many risk management disciplines, such as Business Continuity and Security Risk Management, as set out below:

- Recognition or identification of risks
- Ranking or evaluation of risks
    - Responding to significant risks
    - Tolerate
    - Treat
    - Transfer
    - Terminate
- Resourcing controls
- Reaction Planning
- Reporting & monitoring risk performance
- Reviewing the Risk Management framework

There are a number of guidelines and publications regarding Emergency Planning, published by various professional organizations such as ASIS, National Fire Protection Association (NFPA), and the International Association of Emergency Managers (IAEM). There are very few Emergency Management specific standards, and emergency management as a discipline tends to fall under business resilience standards.

In order to avoid, or reduce significant losses to a business, emergency managers should work to identify and anticipate potential risks, hopefully to reduce their probability of occurring. In the event that an emergency does occur, managers should have a plan prepared to mitigate the effects of that emergency, as well as to ensure Business Continuity of critical operations post-incident. It is essential for an organisation to include procedures for determining whether an emergency situation has occurred and at what point an emergency management plan should be activated.

Implementation Ideals

An emergency plan must be regularly maintained, in a structured and methodical manner, to ensure it is up-to-date in the event of an emergency. Emergency managers generally follow a common process to anticipate, assess, prevent, prepare, respond and recover from an incident.

Pre-incident Training and Testing

Emergency management plans and procedures should include the identification of appropriately trained staff members responsible for decision-making when an emergency occurs. Training plans should include internal people, contractors and civil protection partners, and should state the nature and frequency of training and testing.

A team of emergency responders performs a training scenario involving anthrax.

Testing of a plan's effectiveness should occur regularly. In instances where several business or organisations occupy the same space, joint emergency plans, formally agreed to by all parties, should be put into place.

Communicating and Incident Assessment

Communication is one of the key issues during any emergency, pre-planning of communications is critical. Miscommunication can easily result in emergency events escalating unnecessarily.

Once an emergency has been identified a comprehensive assessment evaluating the level of impact and its financial implications should be undertaken. Following assessment, the appropriate plan or response to be activated will depend on a specific pre-set criteria within the emergency plan. The steps necessary should be prioritized to ensure critical functions are operational as soon as possible.

Phases and Personal Activities

Emergency management consists of five phases: prevention, mitigation, preparedness, response and recovery.

Prevention

It focuses on preventing the human hazard, primarily from potential natural disasters or terrorist attacks. Preventive measures are taken on both the domestic and international levels, designed to provide permanent protection from disasters. Not all disasters, particularly natural disasters, can be prevented, but the risk of loss of life and injury can be mitigated with good evacuation plans, environmental planning and design standards. In January 2005, 167 Governments adopted a 10-year global plan for natural disaster risk reduction called the Hyogo Framework.

Preventing or reducing the impacts of disasters on our communities is a key focus for emergency management efforts today. Prevention and mitigation also help reduce the financial costs of disaster response and recovery. Public Safety Canada is working with provincial and territorial governments and stakeholders to promote disaster prevention and mitigation using a risk-based and all-hazards approach. In 2008, Federal/Provincial/Territorial Ministers endorsed a National Disaster Mitigation Strategy.

Mitigation

Preventive or mitigation measures take different forms for different types of disasters. In earthquake prone areas, these preventive measures might include structural changes such as the installation of an earthquake valve to instantly shut off the natural gas supply, seismic retrofits of property, and the securing of items inside a building. The latter may include the mounting of furniture, refrigerators, water heaters and breakables to the walls, and the addition of cabinet latches. In flood prone areas, houses can be built on poles/stilts. In areas prone to prolonged electricity black-outs installation of a generator ensures continuation of electrical service. The construction of storm cellars and fallout shelters are further examples of personal mitigative actions.

On a national level, governments might implement large scale mitigation measures. After the monsoon floods of 2010, the Punjab government subsequently constructed 22 'disaster-resilient' model villages, comprising 1885 single-storey homes, together with schools and health centres.

Disaster mitigation measures are those that eliminate or reduce the impacts and risks of hazards through proactive measures taken before an emergency or disaster occurs.

One of the best known examples of investment in disaster mitigation is the Red River Floodway. The building of the Floodway was a joint provincial/federal undertaking to protect the City of Winnipeg and reduce the impact of flooding in the Red River Basin. It cost $62.7 million to build in the 1960s. Since then, the floodway has been used over 20 times. Its use during the 1997 Red River Flood alone saved an estimated $4.5 billion in costs from potential damage to the city. The Floodway was expanded in 2006 as a joint provincial/federal initiative.

Preparedness

Preparedness focuses on preparing equipment and procedures for use when a disaster occurs. This equipment and these procedures can be used to reduce vulnerability to disaster, to mitigate the impacts of a disaster or to respond more efficiently in an emergency. The Federal Emergency Management Agency (FEMA) has set out a basic four-stage vision of preparedness flowing from mitigation to preparedness to response to recovery and back to mitigation in a circular planning process. This circular, overlapping model has been modified by other agencies, taught in emergency class and discussed in academic papers. FEMA also operates a Building Science Branch that develops and produces multi-hazard mitigation guidance that focuses on creating disaster-resilient communities to reduce loss of life and property. FEMA advises citizens to prepare their homes with some emergency essentials in the case that the food distribution lines are interrupted. FEMA has subsequently prepared for this contingency by purchasing hundreds of thousands of freeze dried food emergency meals ready to eat (MRE's) to dispense to the communities where emergency shelter and evacuations are implemented.

An airport emergency preparedness exercise

Emergency preparedness can be difficult to measure. CDC focuses on evaluating the effectiveness of its public health efforts through a variety of measurement and assessment programs.

Local Emergency Planning Committees

Local Emergency Planning Committees (LEPCs) are required by the United States Environmental Protection Agency under the Emergency Planning and Community Right-to-Know Act to develop an emergency response plan, review the plan at least annually, and provide information about chemicals in the community to local citizens. This emergency preparedness effort focuses on hazards presented by use and storage of extremely hazardous, hazardous and toxic chemicals. Particular requirements of LEPCs include

- Identification of facilities and transportation routes of extremely hazardous substances
- Description of emergency response procedures, on and off site
- Designation of a community coordinator and facility emergency coordinator(s) to implement the plan
- Outline of emergency notification procedures
- Description of how to determine the probable affected area and population by releases
- Description of local emergency equipment and facilities and the persons responsible for them
- Outline of evacuation plans

- A training program for emergency responders (including schedules)
- Methods and schedules for exercising emergency response plans

According to the EPA, "Many LEPCs have expanded their activities beyond the requirements of EPCRA, encouraging accident prevention and risk reduction, and addressing homeland security in their communities" and the Agency offers advice on how to evaluate the effectiveness of these committees.

Preparedness Measures

Preparedness measures can take many forms ranging from focusing on individual people, locations or incidents to broader, government-based "all hazard" planning. There are a number of preparedness stages between "all hazard' and individual planning, generally involving some combination of both mitigation and response planning. Business continuity planning encourages businesses to have a Disaster Recovery Plan. Community- and faith-based organizations mitigation efforts promote field response teams and inter-agency planning.

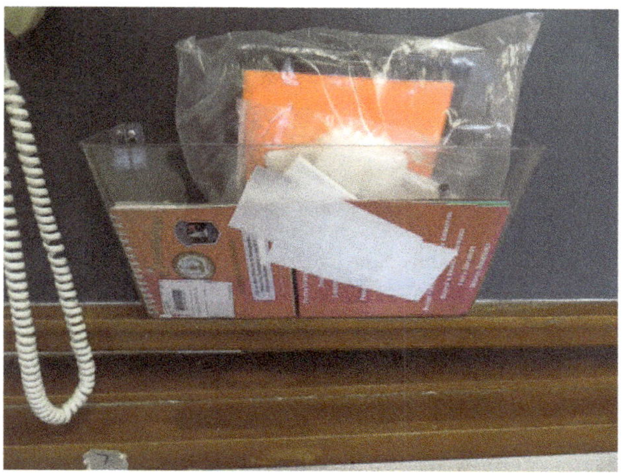

Classroom response kit

School-based response teams cover everything from live shooters to gas leaks and nearby bank robberies. Educational institutions plan for cyberattacks and windstorms. Industry specific guidance exists for horse farms, boat owners and more.

Family preparedness for disaster is fairly unusual. A 2013 survey found that only 19% of American families felt that they were "very prepared" for a disaster. Still, there are many resources available for family disaster planning. The Department of Homeland Security's Ready.gov page includes a Family Emergency Plan Checklist, has a whole webpage devoted to readiness for kids, complete with cartoon-style superheroes, and ran a Thunderclap Campaign in 2014. The Center for Disease Control has a Zombie Apocalypse website.

Disasters take a variety of forms to include earthquakes, tsunamis or regular structure fires. That a disaster or emergency is not large scale in terms of population or acreage impacted or duration does not make it any less of a disaster for the people or area impacted and much can be learned about preparedness from so-called small disasters. The Red Cross states that it responds to nearly 70,000 disasters a year, the most common of which is a single family fire.

Kitchen fire extinguisher

Items on shelves in basement

Preparedness starts with an individual's everyday life and involves items and training that would be useful in an emergency. What is useful in an emergency is often also useful in everyday life. From personal preparedness, preparedness continues on a continuum through family preparedness, community preparedness and then business, non-profit and governmental preparedness. Some organizations blend these various levels. For example, the International Red Cross and Red Crescent Movement has a webpage on disaster training as well as offering training on basic preparedness such as Cardiopulmonary resuscitation and First Aid. Other non-profits such as Team Rubicon bring specific groups of people into disaster preparedness and response operations. FEMA breaks down preparedness into a pyramid, with citizens on the foundational bottom, on top of which rests local government, state government and federal government in that order.

Non-perishable food in cabinet

The basic theme behind preparedness is to be ready for an emergency and there are a number of different variations of being ready based on an assessment of what sort of threats exist. Nonetheless, there is basic guidance for preparedness that is common despite an area's specific dangers. FEMA recommends that everyone have a three-day survival kit for their household. Because individual household sizes and specific needs might vary, FEMA's recommendations are not item specific, but the list includes:

- Three-day supply of non-perishable food.
- Three-day supply of water – one gallon of water per person, per day.
- Portable, battery-powered radio or television and extra batteries.
- Flashlight and extra batteries.
- First aid kit and manual.
- Sanitation and hygiene items (e.g. toilet paper, menstrual hygiene products).
- Matches and waterproof container.
- Whistle.
- Extra clothing.
- Kitchen accessories and cooking utensils, including a can opener.
- Photocopies of credit and identification cards.
- Cash and coins.
- Special needs items, such as prescription medications, eyeglasses, contact lens solutions, and hearing aid batteries.

- Items for infants, such as formula, diapers, bottles, and pacifiers.
- Other items to meet unique family needs.

Along similar lines, but not exactly the same, CDC has its own list for a proper disaster supply kit.

- Water—one gallon per person, per day
- Food—nonperishable, easy-to-prepare items
- Flashlight
- Battery powered or hand crank radio (NOAA Weather Radio, if possible)
- Extra batteries
- First aid kit
- Medications (7-day supply), other medical supplies, and medical paperwork (e.g., medication list and pertinent medical information)
- Multipurpose tool (e.g., Swiss army knife)
- Sanitation and personal hygiene items
- Copies of personal documents (e.g., proof of address, deed/lease to home, passports, birth certificates, and insurance policies)
- Cell phone with chargers
- Family and emergency contact information
- Extra cash
- Emergency blanket
- Map(s) of the area
- Extra set of car keys and house keys
- Manual can opener

Children are a special population when considering Emergency preparedness and many resources are directly focused on supporting them. SAMHSA has list of tips for talking to children during infectious disease outbreaks, to include being a good listener, encouraging children to ask questions and modeling self-care by setting routines, eating healthy meals, getting enough sleep and taking deep breaths to handle stress. FEMA has similar advice, noting that "Disasters can leave children feeling frightened, confused, and insecure" whether a child has experienced it first hand, had it happen to a friend or simply saw it on television. In the same publication, FEMA further notes, "Preparing for disaster helps everyone in the family accept the fact that disasters do happen, and provides an opportunity to identify and collect the resources needed to meet basic needs after disaster. Preparation helps; when people feel prepared, they cope better and so do children."

To help people assess what threats might be in order to augment their emergency supplies or improve their disaster response skills, FEMA has published a booklet called the "Threat and Hazard Identification and Risk Assessment Guide." (THIRA) This guide, which outlines the THIRA process, emphasizes "whole community involvement," not just governmental agencies, in preparedness efforts. In this guide, FEMA breaks down hazards into three categories: Natural, technological and human caused and notes that each hazard should be assessed for both its likelihood and its significance. According to FEMA, "Communities should consider only those threats and hazards that could plausibly occur" and "Communities should consider only those threats and hazards that would have a significant effect on them." To develop threat and hazard context descriptions, communities should take into account the time, place, and conditions in which threats or hazards might occur.

Not all preparedness efforts and discussions involve the government or established NGOs like the Red Cross. Emergency preparation discussions are active on the internet, with many blogs and websites dedicated to discussing various aspects of preparedness. On-line sales of items such as survival food, medical supplies and heirloom seeds allow people to stock basements with cases of food and drinks with 25 year shelf lives, sophisticated medical kits and seeds that are guaranteed to sprout even after years of storage.

One group of people who put a lot of effort in disaster preparations is called Doomsday Preppers. This subset of preparedness-minded people often share a belief that the FEMA or Red Cross emergency preparation suggestions and training are not extensive enough. Sometimes called survivalists, Doomsday Preppers are often preparing for The End Of The World As We Know It, abbreviated as TEOTWAWKI. With a motto some have that "The Future Belongs to those who Prepare," this Preparedness subset has its own set of Murphy's Rules, including "Rule Number 1: Food, you still don't have enough" and "Rule Number 26: People who thought the Government would save them, found out that it didn't."

Not all emergency preparation efforts revolve around food, guns and shelters, though these items help address the needs in the bottom two sections of Maslow's hierarchy of needs. The American Preppers Network has an extensive list of items that might be useful in less apparent ways than a first aid kid or help add 'fun' to challenging times. These items include:

- Books and magazines
- Arts and crafts painting
- Children's entertainment
- Crayons and coloring books
- Notebooks and writing supplies
- Nuts, bolts, screws, nails, etc.
- Religious material
- Sporting equipment, card games and board games
- Posters and banners creating awareness

Emergency preparedness goes beyond immediate family members. For many people, pets are an integral part of their families and emergency preparation advice includes them as well. It is not unknown for pet owners to die while trying to rescue their pets from a fire or from drowning. CDC's Disaster Supply Checklist for Pets includes:

- Food and water for at least 3 days for each pet; bowls, and a manual can opener.
- Depending on the pet you may need a litter box, paper towels, plastic trash bags, grooming items, and/or household bleach.
- Medications and medical records stored in a waterproof container.
- First aid kit with a pet first aid book.
- Sturdy leash, harness, and carrier to transport pet safely. A carrier should be large enough for the animal to stand comfortably, turn around, and lie down. Your pet may have to stay in the carrier for several hours.
- Pet toys and the pet's bed, if you can easily take it, to reduce stress.
- Current photos and descriptions of your pets to help others identify them in case you and your pets become separated, and to prove that they are yours.
- Information on feeding schedules, medical conditions, behavior problems, and the name and telephone number of your veterinarian in case you have to board your pets or place them in foster care.

Emergency preparedness also includes more than physical items and skill-specific training. Psychological preparedness is also a type of emergency preparedness and specific mental health preparedness resources are offered for mental health professionals by organizations such as the Red Cross. These mental health preparedness resources are designed to support both community members affected by a disaster and the disaster workers serving them. CDC has a website devoted to coping with a disaster or traumatic event. After such an event, the CDC, through the Substance Abuse and Mental Health Services Administration (SAMHSA), suggests that people seek psychological help when they exhibit symptoms such as excessive worry, crying frequently, an increase in irritability, anger, and frequent arguing, wanting to be alone most of the time, feeling anxious or fearful, overwhelmed by sadness, confused, having trouble thinking clearly and concentrating, and difficulty making decisions, increased alcohol and/or substance use, increased physical (aches, pains) complaints such as headaches and trouble with "nerves."

Sometimes emergency supplies are kept in what is called a Bug-out bag. While FEMA does not actually use the term "Bug out bag," calling it instead some variation of a "Go Kit," the idea of having emergency items in a quickly accessible place is common to both FEMA and CDC, though on-line discussions of what items a "bug out bag" should include sometimes cover items such as firearms and great knives that are not specifically suggested by FEMA or CDC. The theory behind a "bug out bag" is that emergency preparations should include the possibility of Emergency evacuation. Whether fleeing a burning building or hastily packing a car to escape an impending hurricane, flood or dangerous chemical release, rapid departure from a home or workplace environment is always a possibility and FEMA suggests having a Family Emergency Plan for such occasions. Because family members may not be together when disaster strikes, this plan should include reliable contact information for friends or relatives who live outside of what would be the disaster area for household members to notify they are safe or otherwise communicate with each other. Along with the contact information, FEMA suggests having well-understood local gathering points if a house must be evacuated quickly to avoid the dangers of re-reentering a burning home. Family and emergency contact information should be printed on cards and put in each family member's backpack or wallet. If family members spend a significant amount of time in a specific location, such as at work or school, FEMA suggests learning the emergency preparation plans for those places. FEMA has a specific form, in English and in Spanish, to help people put together these emergency plans, though it lacks lines for email contact information.

Like children, people with disabilities and other special needs have special emergency preparation needs. While "disability" has a specific meaning for specific organizations such as collecting Social Security benefits, for the purposes of emergency preparedness, the Red Cross uses the term in a broader sense to include people with physical, medical, sensor or cognitive disabilities or the elderly and other special needs populations. Depending on the particular disability, specific emergency preparations might be required. FEMA's suggestions for people with disabilities includes having copies of prescriptions, charging devices for medical devices such as motorized wheel chairs and a week's supply of medication readily available LINK or in a "go stay kit." In some instances, lack of competency in English may lead to special preparation requirements and communication efforts for both individuals and responders.

FEMA notes that long term power outages can cause damage beyond the original disaster that can be mitigated with emergency generators or other power sources to provide an Emergency power system. The United States Department of Energy states that 'homeowners, business owners, and local leaders may have to take an active role in dealing with energy disruptions on their own." This active role may include installing or other procuring generators that are either portable or permanently mounted and run on fuels such as propane or natural gas or gasoline. Concerns about carbon monoxide poisoning, electrocution, flooding, fuel storage and fire lead even small property owners to consider professional installation and maintenance. Major institutions like hospitals, military bases and educational institutions often have or are considering extensive backup power systems. Instead of, or in addition to, fuel-based power systems, solar, wind and other alternative power sources may be used. Standalone batteries, large or small, are also used to provide backup charging for electrical systems and devices ranging from emergency lights to computers to cell phones.

Emergency preparedness does not stop at home or at school. The United States Department of Health and Human Services addresses specific emergency preparedness issues hospitals may have to respond to, including maintaining a safe temperature, providing adequate electricity for life support systems and even carrying out evacuations under extreme circumstances. FEMA encourages all businesses to have businesses to have an emergency response plan and the Small Business Administration specifically advises small business owners to also focus emergency preparedness and provides a variety of different worksheets and resources.

FEMA cautions that emergencies happen while people are travelling as well and provides guidance around emergency preparedness for a range travelers to include commuters, *Commuter Emergency Plan* and holiday travelers. In particular, Ready.gov has a number of emergency preparations specifically designed for people with cars. These preparations include having a full gas tank, maintaining adequate windshield wiper fluid and other basic car maintenance tips. Items specific to an emergency include:

- Jumper cables: might want to include flares or reflective triangle
- Flashlights, to include extra batteries (batteries have less power in colder weather)
- First Aid Kit, to include any necessary medications, baby formula and diapers if caring for small children
- Non-perishable food such as canned food (be alert to liquids freezing in colder weather), and protein rich foods like nuts and energy bars
- Manual can opener
- At least 1 gallon of water per person a day for at least 3 days (be alert to hazards of frozen water and resultant container rupture)
- Basic toolkit: pliers, wrench, screwdriver
- Pet supplies: food and water
- Radio: battery or hand cranked
- For snowy areas: cat litter or sand for better tire traction; shovel; ice scraper; warm clothes, gloves, hat, sturdy boots, jacket and an extra change of clothes
- Blankets or sleeping bags
- Charged Cell Phone: and car charger

In addition to emergency supplies and training for various situations, FEMA offers advice on how to mitigate disasters. The Agency gives instructions on how to retrofit a home to minimize hazards from a Flood, to include installing a Backflow prevention device, anchoring fuel tanks and relocating electrical panels.

Marked gas shuttoff

Given the explosive danger posed by natural gas leaks, Ready.gov states unequivocally that "It is vital that all household members know how to shut off natural gas" and that property owners must ensure they have any special tools needed for their particular gas hookups. Ready.gov also notes that "It is wise to teach all responsible household members where and how to shut off the electricity," cautioning that individual circuits should be shut off before the main circuit. Ready.gov further states that "It is vital that all household members learn how to shut off the water at the main house valve" and cautions that the possibility that rusty valves might require replacement.

Response

The response phase of an emergency may commence with Search and Rescue but in all cases the focus will quickly turn to fulfilling the basic humanitarian needs of the affected population. This assistance may be provided by national or international agencies and organizations. Effective coordination of disaster assistance is often crucial, particularly when many organizations respond and local emergency management agency (LEMA) capacity has been exceeded by the demand or diminished by the disaster itself. The National Response Framework is a United States government publication that explains responsibilities and expectations of government officials at the local, state, federal, and tribal levels. It provides guidance on Emergency Support Functions that may be integrated in whole or parts to aid in the response and recovery process.

On a personal level the response can take the shape either of a *shelter in place* or an *evacuation.*

Evacuation sign

In a shelter-in-place scenario, a family would be prepared to fend for themselves in their home for many days without any form of outside support. In an *evacuation*, a family leaves the area by automobile or other mode of transportation, taking with them the maximum amount of supplies they can carry, possibly including a tent for shelter. If mechanical transportation is not available, evacuation on foot would ideally include carrying at least three days of supplies and rain-tight bedding, a tarpaulin and a bedroll of blankets.

Donations are often sought during this period, especially for large disasters that overwhelm local capacity. Due to efficiencies of scale, money is often the most cost-effective donation if fraud is avoided. Money is also the most flexible, and if goods are sourced locally then transportation is minimized and the local economy is boosted. Some donors prefer to send gifts in kind, however these items can end up creating issues, rather than helping. One innovation by Occupy Sandy volunteers is to use a donation registry, where families and businesses impacted by the disaster can make specific requests, which remote donors can purchase directly via a web site.

Medical considerations will vary greatly based on the type of disaster and secondary effects. Survivors may sustain a multitude of injuries to include lacerations, burns, near drowning, or crush syndrome.

Recovery

The recovery phase starts after the immediate threat to human life has subsided. The immediate goal of the recovery phase is to bring the affected area back to normalcy as quickly as possible. During reconstruction it is recommended to consider the location or construction material of the property.

The most extreme home confinement scenarios include war, famine and severe epidemics and may last a year or more. Then recovery will take place inside the home. Planners for these events usually buy bulk foods and appropriate storage and preparation equipment, and eat the food as part of normal life. A simple balanced diet can be constructed from vitamin pills, whole-meal wheat, beans, dried milk, corn, and cooking oil. One should add vegetables, fruits, spices and meats, both prepared and fresh-gardened, when possible.

As a Profession

Professional emergency managers can focus on government and community preparedness, or private business preparedness. Training is provided by local, state, federal and private organizations and ranges from public information and media relations to high-level incident command and tactical skills.

In the past, the field of emergency management has been populated mostly by people with a military or first responder background. Currently, the field has become more diverse, with many managers coming from a variety of backgrounds other than the military or first responder fields. Educational opportunities are increasing for those seeking undergraduate and graduate degrees in emergency management or a related field. There are over 180 schools in the US with emergency management-related programs, but only one doctoral program specifically in emergency management.

Professional certifications such as Certified Emergency Manager (CEM) and Certified Business Continuity Professional (CBCP) are becoming more common as professional standards are raised throughout the field, particularly in the United States. There are also professional organizations for emergency managers, such as the National Emergency Management Association and the International Association of Emergency Managers.

Principles

In 2007, Dr. Wayne Blanchard of FEMA's Emergency Management Higher Education Project, at the direction of Dr. Cortez Lawrence, Superintendent of FEMA's Emergency Management Institute, convened a working group of emergency management practitioners and academics to consider principles of emergency management. This was the first time the principles of the discipline were to be codified. The group agreed on eight principles that will be used to guide the development of a doctrine of emergency management. Below is a summary:

1. Comprehensive – consider and take into account all hazards, all phases, all stakeholders and all impacts relevant to disasters.
2. Progressive – anticipate future disasters and take preventive and preparatory measures to build disaster-resistant and disaster-resilient communities.
3. Risk-driven – use sound risk management principles (hazard identification, risk analysis, and impact analysis) in assigning priorities and resources.
4. Integrated – ensure unity of effort among all levels of government and all elements of a community.
5. Collaborative – create and sustain broad and sincere relationships among individuals and organizations to encourage trust, advocate a team atmosphere, build consensus, and facilitate communication.
6. Coordinated – synchronize the activities of all relevant stakeholders to achieve a common purpose.
7. Flexible – use creative and innovative approaches in solving disaster challenges.
8. Professional – value a science and knowledge-based approach; based on education, training, experience, ethical practice, public stewardship and continuous improvement.

Tools

In recent years the continuity feature of emergency management has resulted in a new concept, Emergency Management Information Systems (EMIS). For continuity and inter-operability between emergency management stakeholders, EMIS supports an infrastructure that integrates emergency plans at all levels of government and non-government involvement for all four phases of emergencies. In the healthcare field, hospitals utilize the Hospital Incident Command System (HICS), which provides structure and organization in a clearly defined chain of command.

Disaster Response Technologies

Smart Emergency Response System (SERS) prototype was built in the SmartAmerica Challenge 2013-2014, a United States government initiative. SERS has been created by a team of nine organizations led by MathWorks. The project was featured at the White House in June 2014 and described by Todd Park (U.S. Chief Technology Officer) as an exemplary achievement.

The Smart America initiative challenges the participants to build cyber-physical systems as a glimpse of the future to save lives, create jobs, foster businesses, and improve the economy. SERS primarily saves lives. The system provides the survivors and the emergency personnel with information to locate and assist each other during a disaster. SERS allows to submit help requests to a MATLAB-based mission center connecting first responders, apps, search-and-rescue dogs, a 6-feet-tall humanoid, robots, drones, and autonomous aircraft and ground vehicles. The command and control center optimizes the available resources to serve every incoming requests and generates an action plan for the mission. The Wi-Fi network is created on the fly by the drones equipped with antennas. In addition, the autonomous rotorcrafts, planes, and ground vehicles are simulated with Simulink and visualized in a 3D environment (Google Earth) to unlock the ability to observe the operations on a mass scale.

Within other Professions

A disaster plan book at Rockefeller University in a biochemistry research laboratory.

Practitioners in emergency management come from an increasing variety of backgrounds. Professionals from memory institutions (e.g., museums, historical societies, etc.) are dedicated to preserving cultural heritage—objects and records. This has been an increasingly major component within this field as a result of the heightened awareness following the September 11 attacks in 2001, the hurricanes in 2005, and the collapse of the Cologne Archives.

To increase the potential successful recovery of valuable records, a well-established and thoroughly tested plan must be developed. This plan should emphasize simplicity in order to aid in response and recovery: employees should perform similar tasks in the response and recovery phase that they perform under normal conditions. It should also include mitigation strategies such as the installation of sprinklers within the institution. Professional associations hold regular workshops to keep individuals up to date with tools and resources in order to minimize risk and maximize recovery.

Other Tools

In 2008, the U.S. Agency for International Development created a web-based tool for estimating populations impacted by disasters. Called Population Explorer the tool uses land scan population data, developed by Oak Ridge National Laboratory, to distribute population at a resolution 1 km^2 for all countries in the world. Used by USAID's FEWS NET Project to estimate populations vulnerable and or imd by food insecurity, Population Explorer is gaining wide use in a range of emergency analysis and response actions, including estimating populations impacted by floods in Central America and the Pacific Ocean Tsunami event in 2009.

In 2007, a checklist for veterinarians was published in the Journal of the American Veterinary Medical Association, it had two sets of questions for a professional to ask themselves before assisting with an emergency:

Absolute Requirements for Participation:

- Have I chosen to participate?
- Have I taken ICS training?
- Have I taken other required background courses?
- Have I made arrangements with my practice to deploy?
- Have I made arrangements with my family?

Incident Participation:

- Have I been invited to participate
- Are my skill sets a match for the mission?
- Can I access just-in-time training to refresh skills or acquire needed new skills?
- Is this a self-support mission?
- Do I have supplies needed for three to five days of self-support?

While written for veterinarians, this checklist is applicable for any professional to consider before assisting with an emergency.

International Organizations

The International Emergency Management Society

The International Emergency Management Society (TIEMS), is an international non-profit NGO, registered in Belgium. TIEMS is a Global Forum for Education, Training, Certification and Policy in Emergency and Disaster Management. TIEMS' goal is to develop and bring modern emergency management tools, and techniques into practice, through the exchange of information, methodology innovations and new technologies.

TIEMS provides a platform for stakeholders to meet, network and learn about new technical and operational methodologies. TIEMS focuses on cultural differences to be understood and included in the society's events, education and research programs. This is achieved by establishing local chapters worldwide. Today, TIEMS has chapters in Benelux, Romania, Finland, Italy, Middle East and North Africa (MENA), Iraq, India, Korea, Japan and China.

International Association of Emergency Managers

The International Association of Emergency Managers (IAEM) is a non-profit educational organization aimed at promoting the goals of saving lives and property protection during emergencies. The mission of IAEM is to serve its members by providing information, networking and professional opportunities, and to advance the emergency management profession.

It has seven councils around the world: Asia, Canada, Europa, International, Oceania, Student and USA.

The Air Force Emergency Management Association, affiliated by membership with the IAEM, provides emergency management information and networking for U.S. Air Force Emergency Management personnel.

International Recovery Platform

The International Recovery Platform (IRP) was conceived at the World Conference on Disaster Reduction (WCDR) in Kobe, Hyogo, Japan in January 2005, as part of the *Hyogo Framework for Action (HFA) 2005–2015*. The HFA is a global plan for disaster risk reduction adopted by 168 governments.

The key role of IRP is to identify gaps in post disaster recovery and to serve as a catalyst for the development of tools and resources for recovery efforts.

The International Red Cross and Red Crescent Movement

The International Federation of Red Cross and Red Crescent Societies (IFRC) works closely with National Red Cross and Red Crescent societies in responding to emergencies, many times playing a pivotal role. In addition, the IFRC may deploy assessment teams, e.g. Field Assessment and Coordination Teams (FACT), to the affected country if requested by the national society. After assessing the needs, Emergency Response Units (ERUs) may be deployed to the affected country or region. They are specialized in the response component of the emergency management framework.

Baptist Global Response

Baptist Global Response (BGR) is a disaster relief and community development organization. BGR and its partners respond globally to people with critical needs worldwide, whether those needs arise from chronic conditions or acute crises such as natural disasters. While BGR is not an official entity of the Southern Baptist Convention, it is rooted in Southern Baptist life and is the international partnership of Southern Baptist Disaster Relief teams, which operate primarily in the US and Canada.

United Nations

The United Nations system rests with the Resident Coordinator within the affected country. However, in practice, the UN response will be coordinated by the UN Office for the Coordination of Humanitarian Affairs (UN-OCHA), by deploying a UN Disaster Assessment and Coordination (UNDAC) team, in response to a request by the affected country's government.

World Bank

Since 1980, the World Bank has approved more than 500 projects related to disaster management, dealing with both disaster mitigation as well as reconstruction projects, amounting to more than US$40 billion. These projects have taken place all over the world, in countries such as Argentina, Bangladesh, Colombia, Haiti, India, Mexico, Turkey and Vietnam.

Prevention and mitigation projects include forest fire prevention measures, such as early warning measures and education campaigns; early-warning systems for hurricanes; flood prevention mechanisms (e.g. shore protection, terracing, etc.); and earthquake-prone construction. In a joint venture with Columbia University under the umbrella of the ProVention Consortium the World Bank has established a Global Risk Analysis of Natural Disaster Hotspots.

In June 2006, the World Bank, in response to the HFA, established the Global Facility for Disaster Reduction and Recovery (GFDRR), a partnership with other aid donors to reduce disaster losses. GFDRR helps developing countries fund development projects and programs that enhance local capacities for disaster prevention and emergency preparedness.

European Union

In 2001 the EU adopted Community Mechanism for Civil Protection, to facilitate co-operation in the event of major emergencies requiring urgent response actions. This also applies to situations where there may be an imminent threat as well.

The heart of the Mechanism is the Monitoring and Information Center (MIC), part of the European Commission's Directorate-General for Humanitarian Aid & Civil Protection. Accessible 24 hours a day, it gives countries access to a one-stop-shop of civil protections available amongst all the participating states. Any country inside or outside the Union affected by a major disaster can make an appeal for assistance through the MIC. It acts as a communication hub, and provides useful and updated information on the actual status of an ongoing emergency.

National Organizations

Australia

Natural disasters are part of life in Australia. Heatwaves have killed more Australians than any other type of natural disaster in the 20th century. Australia's emergency management processes embrace the concept of the prepared community. The principal government agency in achieving this is Emergency Management Australia.

Canada

Public Safety Canada is Canada's national emergency management agency. Each province is required to have both legislation for dealing with emergencies, and provincial emergency management agencies, typically called "Emergency Measures Organizations" (EMO). Public Safety Canada co-ordinates and supports the efforts of federal organizations as well as other levels of government, first responders, community groups, the private sector, and other nations. The Public Safety and Emergency Preparedness Act defines the powers, duties and functions of PS are outlined. Other acts are specific to individual fields such as corrections, law enforcement, and national security.

Germany

In Germany the Federal Government controls the German *Katastrophenschutz* (disaster relief), the Technisches Hilfswerk (*Federal Agency for Technical Relief*, THW), and the *Zivilschutz* (civil protection) programs coordinated by the *Federal Office of Civil Protection and Disaster Assistance*. Local fire department units, the German Armed Forces (Bundeswehr), the German Federal Police and the 16 state police forces (Länderpolizei) are also deployed during disaster relief operations.

There are several private organizations in Germany that also deal with emergency relief. Among these are the German Red Cross, Johanniter-Unfall-Hilfe (the German equivalent of the St. John Ambulance), the Malteser-Hilfsdienst, and the Arbeiter-Samariter-Bund. As of 2006, there is a program of study at the University of Bonn leading to the degree "Master in Disaster Prevention and Risk Governance" As a support function radio amateurs provide additional emergency communication networks with frequent trainings.

India

A protective wall built on the shore of the coastal town of Kalpakkam, in aftermath of the 2004 Indian Ocean earthquake.

The National Disaster Management Authority is the primary government agency responsible for planning and capacity-building for disaster relief. Its emphasis is primarily on strategic risk management and mitigation, as well as developing policies and planning. The National Institute of Disaster Management is a policy think-tank and training institution for developing guidelines and training programs for mitigating disasters and managing crisis response.

The National Disaster Response Force is the government agency primarily responsible for emergency management during natural and man-made disasters, with specialized skills in search, rescue and rehabilitation. The Ministry of Science and Technology also contains an agency that brings the expertise of earth scientists and meteorologists to emergency management. The Indian Armed Forces also plays an important role in the rescue/recovery operations after disasters.

Aniruddha's Academy of Disaster Management (ACDM) is a non-profit organization in Mumbai, India with 'disaster management' as its principal objective.

Malaysia

In Malaysia, The National Security Council has the responsibility to handle emergency and disaster events. Ministry of Home Affairs Malaysia, Ministry of Health Malaysia and Ministry of Housing, Urban Wellbeing and Local Government Malaysia are also having responsibility in managing emergency. Several agencies are involved in emergency managements are Royal Malaysian Police, Malaysian Fire and Rescue Department, Malaysian Civil Defence Force, Ministry of Health Malaysia and Malaysian Maritime Enforcement Agency. There were also some voluntary organisation who involved themselves in emergency/ disaster management such as St. John Ambulance of Malaysia, Malaysian Red Crescent Society and so on.

New Zealand

In New Zealand, depending on the scope of the emergency/disaster, responsibility may be handled at either the local or national level. Within each region, local governments are organized into 16 Civil Defence Emergency Management Groups (CMGs). If local arrangements are overwhelmed, pre-existing mutual-support arrangements are activated. Central government has the authority to coordinate the response through the National Crisis Management Centre (NCMC), operated by the Ministry of Civil Defence & Emergency Management (MCDEM). These structures are defined by regulation, and explained in *The Guide to the National Civil Defence Emergency Management Plan 2006*, roughly equivalent to the U.S. Federal Emergency Management Agency's National Response Framework.

New Zealand uses unique terminology for emergency management. Emergency management is rarely used, many government publications retaining the use of the term civil defence. For example, the Minister of Civil Defence is responsible for the MCDEM. Civil Defence Emergency Management is a term in its own right, defined by statute. And disaster rarely appears in official publications, emergency and incident being the preferred terms, with the term event also being used. For example, publications refer to the Canterbury Snow Event 2002.

"4Rs" is the emergency management cycle used in New Zealand, its four phases are known as:

- Reduction = Mitigation
- Readiness = Preparedness
- Response
- Recovery

Pakistan

Disaster management in Pakistan revolves around flood disasters focusing on rescue and relief.

There is a dearth of knowledge and information about hazard identification, risk assessment and management, and disaster preparedness. Disaster management, development planning and environmental management institutions operate in isolation with no integrated planning, there being no central authority for integrated disaster management. State-level measures are heavily tilted towards structural aspects.

Russia

In Russia, the Ministry of Emergency Situations (EMERCOM) is engaged in fire fighting, civil defense, and search and rescue after both natural and human-made disasters.

Somalia

In Somalia, the Federal Government announced in May 2013 that the Cabinet had approved draft legislation on a new Somali Disaster Management Agency (SDMA), which had originally been proposed by the Ministry of Interior. According to the Prime Minister's Media Office, the SDMA will lead and coordinate the government's response to various natural disasters. It is part of a broader effort by the federal authorities to re-establish national institutions. The Federal Parliament is now expected to deliberate on the proposed bill for endorsement after any amendments.

The Netherlands

In the Netherlands the Ministry of Security and Justice is responsible for emergency preparedness and emergency management on a national level and operates a national crisis centre (NCC). The country is divided into 25 safety regions (veiligheidsregio). In a safety region, there are four components: the regional fire department, the regional department for medical care(ambulances and psycho-sociological care etc.), the regional dispatch and a section for risk- and crisis management. The regional dispatch operates for police, fire department and the regional medical care. The dispatch has all these three services combined into one dispatch for the best multi-coordinated response to an incident or an emergency. And also facilitates in information management, emergency communication and care of citizens. These services are the main structure for a response to an emergency. It can happen that, for a specific emergency, the co-operation with an other service is needed, for instance the Ministry of Defence, water board(s) or Rijkswaterstaat. The veiligheidsregio can integrate these other services into their structure by adding them to specific conferences on operational or administrative level.

All regions operate according to the Coordinated Regional Incident Management system.

United Kingdom

Following the 2000 fuel protests and severe flooding that same year, as well as the foot-and-mouth crisis in 2001, the United Kingdom passed the Civil Contingencies Act 2004 (CCA). The CCA defined some organisations as Category 1 and 2 Responders, setting responsibilities regarding emergency preparedness and response. It is managed by the Civil Contingencies Secretariat through Regional Resilience Forums and local authorities.

Disaster Management training is generally conducted at the local level, and consolidated through professional courses that can be taken at the Emergency Planning College. Diplomas, undergraduate and postgraduate qualifications can be gained at universities throughout the country. The Institute of Emergency Management is a charity, established in 1996, providing consulting services for the government, media and commercial sectors. There are a number of professional societies for Emergency Planners including the Emergency Planning Society and the Institute of Civil Protection and Emergency Management.

One of the largest emergency exercises in the UK was carried out on 20 May 2007 near Belfast, Northern Ireland: a simulated plane crash-landing at Belfast International Airport. Staff from five hospitals and three airports participated in the drill, and almost 150 international observers assessed its effectiveness.

United States

Disaster management in the United States has utilized the functional All-Hazards approach for over 20 years, in which managers develop processes (such as communication & warning or sheltering) rather than developing single-hazard or threat focused plans (e.g., a tornado plan). Processes are then mapped to specific hazards or threats, with the manager looking for gaps, overlaps, and conflicts between processes.

Given these notions, emergency managers must identify, contemplate, and assess possible manmade threats and natural threats that may affect their respective locales. Because of geographical differences throughout the nation, a variety of different threats accept communities among the states. Thus, although similarities may exist, no two emergency plans will be completely identical. Additionally, each locale has different resources and capacities (e.g., budgets, personnel, equipment, etc.) for dealing with emergencies. Each individual community must craft its own unique emergency plan that addresses potential threats that are specific to the locality.

This creates a plan more resilient to unique events because all common processes are defined, and encourages planning done by the stakeholders who are closer to the individual processes, such as a traffic management plan written by public works director. This type of planning can lead to conflict with non-emergency management regulatory bodies, which require development of hazard/threat specific plans, such as development of specific H1N1 flu plans and terrorism-specific plans.

In the United States, all disasters are initially local, with local authorities, with usually a police, fire, or EMS agency, taking charge. Many local municipalities may also have a separate dedicated office of emergency management (OEM), along with personnel and equipment. If the event becomes overwhelming to local government, state emergency management (the primary government structure of the United States) becomes the controlling emergency management agency. Federal Emergency Management Agency (FEMA), part of the Department of Homeland Security (DHS), is lead federal agency for emergency management. The United States and its territories are broken down into ten regions for FEMA's emergency management purposes. FEMA supports, but does not override, state authority.

The Citizen Corps is an organization of volunteer service programs, administered locally and coordinated nationally by DHS, which seek to mitigate disasters and prepare the population for emergency response through public education, training, and outreach. Most disaster response is carried out by volunteer organizations. In the US, the Red Cross is chartered by Congress to coordinate disaster response services, including typically being the lead agency handling shelter and feeding of evacuees. Religious organizations, with their ability to provide volunteers quickly, are usually integral during the response process. The largest being the Salvation Army, with a primary focus on chaplaincy and rebuilding, and Southern Baptists who focus on food preparation and distribution, as well as cleaning up after floods and fires, chaplaincy, mobile shower units, chainsaw crews and more. With over 65,000 trained volunteers Southern Baptist Disaster Relief is one of the largest disaster relief organizations in the US. Similar services are also provided by Methodist Relief Services, the Lutherans, and Samaritan's Purse. Unaffiliated volunteers show up at most large disasters. To prevent abuse by criminals and for the safety of the volunteers, procedures have been implemented within most response agencies to manage and effectively use these 'SUVs' (Spontaneous Unaffiliated Volunteers).

The US Congress established the Center for Excellence in Disaster Management and Humanitarian Assistance (COE) as the principal agency to promote disaster preparedness in the Asia-Pacific region.

The National Tribal Emergency Management Council (NEMC) is a non-profit educational organization developed for Tribal organizations to share information and best practices, as well as discussing issues regarding public health and safety, emergency management and homeland security, affecting those under Indian sovereignty. NTMC is organized into Regions, based on the FEMA 10 region system. NTMC was founded by the Northwest Tribal Emergency Management Council (NWTEMC), a consortium of 29 Tribal Nations and Villages in Washington, Idaho, Oregon and Alaska.

If a disaster or emergency is declared to be terror related or an "Incident of National Significance", the Secretary of Homeland Security will initiate the National Response Framework (NRF). The NRF allows the integration of federal resources with local, country, state, or tribal entities, with management of those resources to be handled at the lowest possible level, utilizing the National Incident Management System (NIMS).

FEMA's Emergency Management Institute

Emergency Management Institute's Main Campus in Emmitsburg, Maryland

The Emergency Management Institute (EMI) serves as the national focal point for the development and delivery of emergency management training to enhance the capabilities of state, territorial, local, and tribal government officials; volunteer organizations; FEMA's disaster workforce; other Federal agencies; and the public and private sectors to minimize the impact of disasters and emergencies on the American public. EMI curricula are structured to meet the needs of this diverse audience with an emphasis on separate organizations working together in all-hazards emergencies to save lives and protect property. Particular emphasis is placed on governing doctrine such as the National Response Framework (NRF), National Incident Management System (NIMS), and the National Preparedness Guidelines. EMI is fully accredited by the International Association for Continuing Education and Training (IACET) and the American Council on Education (ACE).

Approximately 5,500 participants attend resident courses each year while 100,000 individuals participate in non-resident programs sponsored by EMI and conducted by state emergency management agencies under cooperative agreements with FEMA. Another 150,000 individuals participate in EMI-supported exercises, and approximately 1,000 individuals participate in the Chemical Stockpile Emergency Preparedness Program (CSEPP).

The *independent study* program at EMI consists of free courses offered to United States citizens in Comprehensive Emergency Management techniques. Course IS-1 is entitled "Emergency Manager: An Orientation to the Position" and provides background information on FEMA and the role of emergency managers in agency and volunteer organization coordination. The EMI Independent Study (IS) Program, a Web-based distance learning program open to the public, delivered extensive online training with approximately 200 courses and trained more than 2.8 million individuals. The EMI IS Web site receives 2.5 to 3 million visitors a day.

2

Understanding Natural Disasters

Natural disasters are disasters that are caused naturally. Some of the examples of natural disasters are earthquakes, sinkholes, limnic eruptions and tsunamis. This section also discusses meteorological disasters such as blizzards, cyclones, droughts and thunderstorms. The topics explained are of great importance to broaden the existing knowledge on natural disasters.

Natural Disaster

A natural disaster is a major adverse event resulting from natural processes of the Earth; examples include floods, hurricanes, tornadoes, volcanic eruptions, earthquakes, tsunamis, and other geologic processes. A natural disaster can cause loss of life or property damage, and typically leaves some economic damage in its wake, the severity of which depends on the affected population's resilience, or ability to recover and also on the infrastructure available.

A rope tornado in its dissipating stage, Tecumseh, Oklahoma.

An adverse event will not rise to the level of a disaster if it occurs in an area without vulnerable population. In a vulnerable area, however, such as Nepal during the 2015 earthquake, an earthquake can have disastrous consequences and leave lasting damage, requiring years to repair.

A daytime wildfire in California.

1755 copper engraving depicting Lisbon in ruins and in flames after the 1755 Lisbon earthquake. A tsunami overwhelms the ships in the harbor.

Geological Disasters

Avalanches and Landslides

During World War I, an estimated 40,000 to 80,000 soldiers died as a result of avalanches during the mountain campaign in the Alps at the Austrian-Italian front. Many of the avalanches were caused by artillery fire.

Earthquakes

An earthquake is the result of a sudden release of energy in the Earth's crust that creates seismic waves. At the Earth's surface, earthquakes manifest themselves by vibration, shaking and sometimes displacement of the ground. Earthquakes are caused by slippage within geological faults. The underground point of origin of the earthquake is called the *seismic focus*. The point directly above the focus on the surface is called the *epicenter*. Earthquakes by themselves rarely kill people or wildlife. It is usually the secondary events that they trigger, such as building collapse, fires, tsunamis (seismic sea waves) and volcanoes, which are actually the human disaster. Many of these could possibly be avoided by better construction, safety systems, early warning and planning.

Sinkholes

The Red Lake (Croatia).

When natural erosion or human mining makes the ground too weak to support the structures built on it, the ground can collapse and produce a sinkhole. For example, the 2010 Guatemala City sinkhole which killed fifteen people was caused when heavy rain from Tropical Storm Agatha, diverted by leaking pipes into a pumice bedrock, led to the sudden collapse of the ground beneath a factory building.

Volcanic Eruptions

Artist's impression of the volcanic eruptions that formed the Deccan Traps in India.

Volcanoes can cause widespread destruction and consequent disaster in several ways. The effects include the volcanic eruption itself that may cause harm following the explosion of the volcano or the fall of rock. Second, lava may be produced during the eruption of a volcano. As it leaves the volcano, the lava destroys many buildings, plants and animals due to its extreme heat . Third, volcanic ash generally meaning the cooled ash - may form a cloud, and settle thickly in nearby locations. When mixed with water this forms a concrete-like material. In sufficient quantity ash may cause roofs to collapse under its weight but even small quantities will harm humans if inhaled. Since the ash has the consistency of ground glass it causes abrasion damage to moving parts such as engines. The main killer of humans in the immediate surroundings of a volcanic eruption is the pyroclastic flows, which consist of a cloud of hot volcanic ash which builds up in the air above the volcano and rushes down the slopes when the eruption no longer supports the lifting of the gases. It is believed that Pompeii was destroyed by a pyroclastic flow. A lahar is a volcanic mudflow or landslide. The 1953 Tangiwai disaster was caused by a lahar, as was the 1985 Armero tragedy in which the town of Armero was buried and an estimated 23,000 people were killed.

A specific type of volcano is the supervolcano. According to the Toba catastrophe theory, 75,000 to 80,000 years ago a supervolcanic event at Lake Toba reduced the human population to 10,000 or even 1,000 breeding pairs, creating a bottleneck in human evolution. It also killed three-quarters of all plant life in the northern hemisphere. The main danger from a supervolcano is the immense cloud of ash, which has a disastrous global effect on climate and temperature for many years.

Hydrological Disasters

The Limpopo River during the 2000 Mozambique flood

It is a violent, sudden and destructive change either in quality of earth's water or in distribution or movement of water on land below the surface or in atmosphere.

Floods

A flood is an overflow of water that 'submerges' land. The EU Floods Directive defines a flood as a temporary covering by water of land which is usually not covered by water. In the sense of 'flowing water', the word may also be applied to the inflow of the tides. Flooding may result from the volume of water within a body of water, such as a river or lake, which overflows causing the result that some of the water escapes its usual boundaries. While the size of a lake or other body of water will vary with seasonal changes in precipitation and snow melt, it is not a significant flood unless the water covers land used by man like a village, city or other inhabited area, roads, expanses of farmland, etc.

Limnic Eruptions

A limnic eruption occurs when a gas, usually CO_2, suddenly erupts from deep lake water, posing the threat of suffocating wildlife, livestock and humans. Such an eruption may also cause tsunamis in the lake as the rising gas displaces water. Scientists believe landslides, volcanic activity, or explosions can trigger such an eruption. To date, only two limnic eruptions have been observed and recorded. In 1984, in Cameroon, a limnic eruption in Lake Monoun caused the deaths of 37 nearby residents, and at nearby Lake Nyos in 1986 a much larger eruption killed between 1,700 and 1,800 people by asphyxiation.

Tsunami

A tsunami also known as a seismic sea wave or as a tidal wave, is a series of waves in a water body caused by the displacement of a large volume of water, generally in an ocean or a large lake. Tsunamis can be caused by undersea earthquakes such as the 2004 Boxing Day tsunami, or by landslides such as the one in 1958 at Lituya Bay, Alaska, or by volcanic eruptions such as the ancient eruption of Santorini. On March 11, 2011, a tsunami occurred near Fukushima, Japan and spread through the Pacific.

Meteorological Disasters

Young steer after a blizzard, March 1966

Blizzards

Blizzards are severe winter storms characterized by heavy snow and strong winds. When high winds stir up snow that has already fallen, it is known as a ground blizzard. Blizzards can impact local economic activities, especially in regions where snowfall is rare. The Great Blizzard of 1888 affected the United States, when many tons of wheat crops were destroyed, and in Asia, 2008 Afghanistan blizzard and the 1972 Iran blizzard were also significant events.

Cyclonic Storms

Cyclone, tropical cyclone, hurricane, and typhoon are different names for the same phenomenon, which is a cyclonic storm system that forms over the oceans. The determining factor on which term is used is based on where they originate. In the Atlantic and Northeast Pacific, the term "hurricane" is used; in the Northwest Pacific it is referred to as a "typhoon" and "cyclones" occur in the South Pacific and Indian Ocean.

The deadliest hurricane ever was the 1970 Bhola cyclone; the deadliest Atlantic hurricane was the Great Hurricane of 1780 which devastated Martinique, St. Eustatius and Barbados. Another notable hurricane is Hurricane Katrina, which devastated the Gulf Coast of the United States in 2005.

Droughts

Drought is the unusual dryness of soil, resulting in crop failure and shortage of water and for other uses which is caused by significant low rainfall than average over a prolonged period. Hot dry winds, shortage of water, high temperatures and consequent evaporation of moisture from the ground can contribute to conditions of drought.

Well-known historical droughts include the 1997–2009 Millennium Drought in Australia led to a water supply crisis across much of the country. As a result, many desalination plants were built for the first time. In 2011, the State of Texas lived under a drought emergency declaration for the entire calendar year and severe economic losses. The drought caused the Bastrop fires.

Thunderstorms

A large hailstone, about 6 cm (2.4 in) in diameter

Severe storms, dust clouds and volcanic eruptions can generate lightning. Apart from the damage typically associated with storms, such as winds, hail and flooding, the lightning itself can damage buildings, ignite fires and kill by direct contact. Especially deadly lightning incidents include a 2007 strike in Ushari Dara, a remote mountain village in northwestern Pakistan, that killed 30 people, the crash of LANSA Flight 508 which killed 91, and a fuel explosion in Dronka, Egypt caused by lightning in 1994 which killed 469. Most lightning deaths occur in the poor countries of America and Asia, where lightning is common and adobe mud brick housing provides little protection.

Hailstorms

Hailstorms are falls of rain drops that arrive as ice, rather than melting before they hit the ground. A particularly damaging hailstorm hit Munich, Germany, on July 12, 1984, causing about 2 billion dollars in insurance claims.

Heat Waves

A heat wave is a period of unusually and excessively hot weather. The worst heat wave in recent history was the European Heat Wave of 2003. A summer heat wave in Victoria, Australia, created conditions which fuelled the massive bushfires in 2009. Melbourne experienced three days in a row of temperatures exceeding 40 °C (104 °F) with some regional areas sweltering through much higher temperatures. The bushfires, collectively known as "Black Saturday", were partly the act of arsonists. The 2010 Northern Hemisphere summer resulted in severe heat waves, which killed over 2,000 people. It resulted in hundreds of wildfires which causing widespread air pollution, and burned thousands of square miles of forest.

A classic anvil-shaped, and clearly-developed Cumulonimbus incus

Tornadoes

A tornado is a violent, dangerous, rotating column of air that is in contact with both the surface of the earth and a cumulonimbus cloud or, in rare cases, the base of a cumulus cloud. It is also referred to as a *twister* or a *cyclone*, although the word cyclone is used in meteorology in a wider sense, to refer to any closed low pressure circulation. Tornadoes come in many shapes and sizes, but are typically in the form of a visible condensation funnel, whose narrow end touches the earth and is often encircled by a cloud of debris and dust. Most tornadoes have wind speeds less than 110 miles per hour (177 km/h), are approximately 250 feet (80 m) across, and travel a few miles (several kilometers) before dissipating. The most extreme tornadoes can attain wind speeds of more than 300 mph (480 km/h), stretch more than two miles (3 km) across, and stay on the ground for dozens of miles (perhaps more than 100 km).

Wildfires

Wildfires are large fires which often start in wildland areas. Common causes include lightning and drought but wildfires may also be started by human negligence or arson. They can spread to populated areas and can thus be a threat to humans and property, as well as wildlife. Notable cases of wildfires were the 1871 Peshtigo Fire in the United States, which killed at least 1700 people, and the 2009 Victorian bushfires in Australia.

Health Disasters

Epidemics

An epidemic is an outbreak of a contractible disease that spreads through a human population. A pandemic is an epidemic that has spread globally. There have been many epidemics throughout history, such as the Black Death. In the last hundred years, significant pandemics include the 1918 Spanish flu pandemic (killing an estimated 50 million people worldwide), the 1957–58 Asian flu pandemic, and the AIDS pandemic, which began in 1959. The 1968–69 Hong Kong water flu pandemic, the 2002-3 SARS pandemic, and the H1N1 Influenza (Swine Flu) Pandemic in 2009–2010 were also significant cases.

Space Disasters

Fallen trees caused by the Tunguska meteoroid of the Tunguska event in June 1908.

Impact Events and Airburst

Asteroids that impact the Earth have led to several major extinction events, including one that created the Chicxulub crater 64.9 million years ago and associated with the demise of the dinosaurs. Scientists estimate that the likelihood of death for a living human from a global impact event is comparable to death from airliner crash.

No human death has been definitively attributed to an impact event, but the 1490 Ch'ing-yang event in which over 10,000 people may have died has been linked to a meteor shower. Even asteroids and comets that burn up in the atmosphere can cause significant destruction on the ground due to the air burst explosion – notable air bursts include the Tunguska event in June 1908, which devastated large areas of Siberian countryside, and the Chelyabinsk meteor on 15 February 2013, which caused widespread property damage in the city of Chelyabinsk and injured 1,491.

Solar Flare

A solar flare is a phenomenon where the sun suddenly releases a great amount of solar radiation, much more than normal. Solar flares are unlikely to cause any direct injury, but can destroy electrical equipment. The potential of solar storms to cause disaster was seen during the 1859 Carrington event, which disrupted the telegraph network, and the March 1989 geomagnetic storm which blacked out Quebec. Some major known solar flares include the X20 event on August 16, 1989, and a similar flare on April 2, 2001. The most powerful flare ever recorded occurred on November 4, 2003 (estimated at between X40 and X45).

Protection by International Law

International law, for example Geneva Conventions defines International Red Cross and Red Crescent Movement the Convention on the Rights of Persons with Disabilities, requires that "States shall take, in accordance with their obligations under international law, including international humanitarian law and international human rights law, all necessary measures to ensure the protection and safety of persons with disabilities in situations of risk, including the occurrence of natural disaster." And further United Nations Office for the Coordination of Humanitarian Affairs is formed by General Assembly Resolution 44/182. People displaced due to natural disasters are currently protected under international law (Guiding Principles of International Displacement, Campala Convention of 2009).

Political Consequences

Natural disasters can also affect political relations with countries and vice versa. Violent conflicts within states can exacerbate the impact of natural disasters by weakening the ability of states, communities and individuals to provide disaster relief. Natural disasters can also worsen on-going conflicts within states by weakening the capacity of states to fight rebels. In developed countries like the US, studies find that incumbents lose votes when the electorate perceives them as responsible for a poor disaster response.

Recent History

In 2012, there were 905 natural disasters worldwide, 93% of which were weather-related disasters. Overall costs were US$170 billion and insured losses $70 billion. 2012 was a moderate year. 45% were meteorological (storms), 36% were hydrological (floods), 12% were climatological (heat waves, cold waves, droughts, wildfires) and 7% were geophysical events (earthquakes and volcanic eruptions). Between 1980 and 2011 geophysical events accounted for 14% of all natural catastrophes.

Studies on natural events requires complete historical records and strategies related to obtaining and storing reliable records, allowing for both critical interpretation and validation of the sources. Under this point of view the irreplaceable role of traditional repositories (archives) can be supplemented by the use of web sources as eBay.

Earthquake

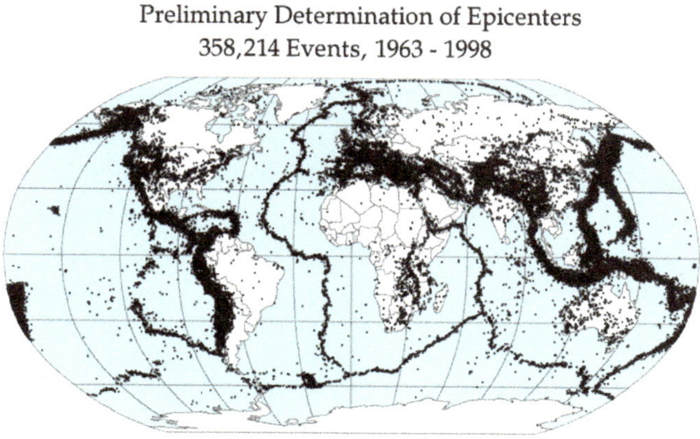

Global earthquake epicenters, 1963–1998

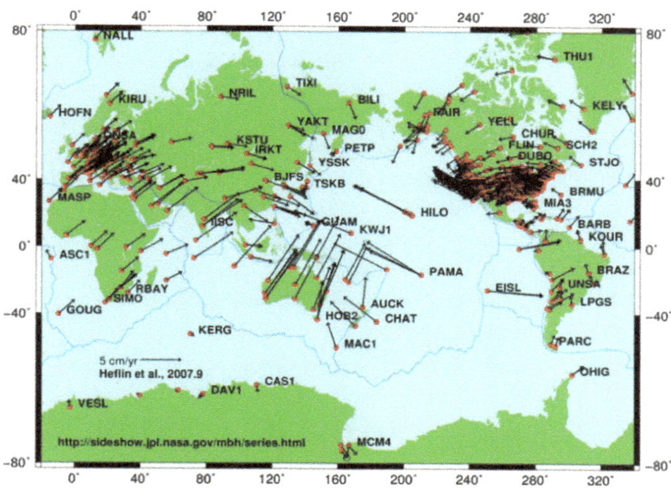

Global plate tectonic movement

An earthquake (also known as a quake, tremor or temblor) is the perceptible shaking of the surface of the Earth, resulting from the sudden release of energy in the Earth's crust that creates seismic waves. Earthquakes can be violent enough to toss people around and destroy whole cities. The seismicity or seismic activity of an area refers to the frequency, type and size of earthquakes experienced over a period of time.

Earthquakes are measured using observations from seismometers. The moment magnitude is the most common scale on which earthquakes larger than approximately 5 are reported for the entire globe. The more numerous earthquakes smaller than magnitude 5 reported by national seismological observatories are measured mostly on the local magnitude scale, also referred to as the Richter magnitude scale. These two scales are numerically similar over their range of validity. Magnitude 3 or lower earthquakes are mostly imperceptible or weak and magnitude 7 and over potentially cause serious damage over larger areas, depending on their depth. The largest earthquakes in historic times have been of magnitude slightly over 9, although there is no limit to the possible magnitude. Intensity of shaking is measured on the modified Mercalli scale. The shallower an earthquake, the more damage to structures it causes, all else being equal.

At the Earth's surface, earthquakes manifest themselves by shaking and sometimes displacement of the ground. When the epicenter of a large earthquake is located offshore, the seabed may be displaced sufficiently to cause a tsunami. Earthquakes can also trigger landslides, and occasionally volcanic activity.

In its most general sense, the word *earthquake* is used to describe any seismic event — whether natural or caused by humans — that generates seismic waves. Earthquakes are caused mostly by rupture of geological faults, but also by other events such as volcanic activity, landslides, mine blasts, and nuclear tests. An earthquake's point of initial rupture is called its focus or hypocenter. The epicenter is the point at ground level directly above the hypocenter.

Naturally Occurring Earthquakes

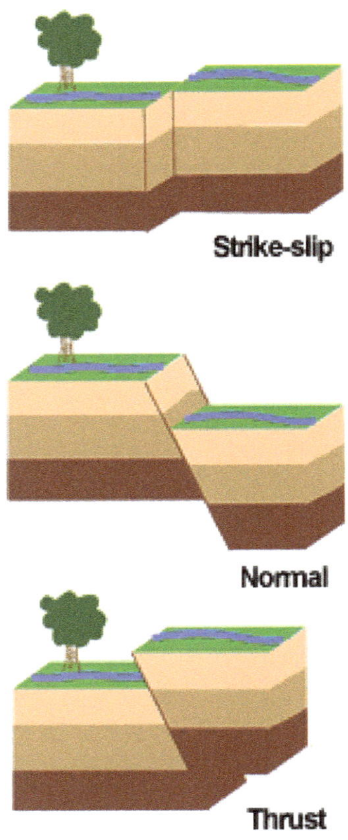

Fault types

Tectonic earthquakes occur anywhere in the earth where there is sufficient stored elastic strain energy to drive fracture propagation along a fault plane. The sides of a fault move past each other smoothly and aseismically only if there are no irregularities or asperities along the fault surface that increase the frictional resistance. Most fault surfaces do have such asperities and this leads to a form of stick-slip behavior. Once the fault has locked, continued relative motion between the plates leads to increasing stress and therefore, stored strain energy in the volume around the fault surface. This continues until the stress has risen sufficiently to break through the asperity, suddenly allowing sliding over the locked portion of the fault, releasing the stored energy. This energy is released as a combination of radiated elastic strain seismic waves, frictional heating of the fault surface, and cracking of the rock, thus causing an earthquake. This process of gradual build-up of strain and stress punctuated by occasional sudden earthquake failure is referred to as the elastic-rebound theory. It is estimated that only 10 percent or less of an earthquake's total energy is radiated as seismic energy. Most of the earthquake's energy is used to power the earthquake fracture growth or is converted into heat generated by friction. Therefore, earthquakes lower the Earth's available elastic potential energy and raise its temperature, though these changes are negligible compared to the conductive and convective flow of heat out from the Earth's deep interior.

Earthquake Fault Types

There are three main types of fault, all of which may cause an interplate earthquake: normal, reverse (thrust) and strike-slip. Normal and reverse faulting are examples of dip-slip, where the displacement along the fault is in the direction of dip and movement on them involves a vertical component. Normal faults occur mainly in areas where the crust is being extended such as a divergent boundary. Reverse faults occur in areas where the crust is being shortened such as at a convergent boundary. Strike-slip faults are steep structures where the two sides of the fault slip horizontally past each other; transform boundaries are a particular type of strike-slip fault. Many earthquakes are caused by movement on faults that have components of both dip-slip and strike-slip; this is known as oblique slip.

Reverse faults, particularly those along convergent plate boundaries are associated with the most powerful earthquakes, megathrust earthquakes, including almost all of those of magnitude 8 or more. Strike-slip faults, particularly continental transforms, can produce major earthquakes up to about magnitude 8. Earthquakes associated with normal faults are generally less than magnitude 7. For every unit increase in magnitude, there is a roughly thirtyfold increase in the energy released. For instance, an earthquake of magnitude 6.0 releases approximately 30 times more energy than a 5.0 magnitude earthquake and a 7.0 magnitude earthquake releases 900 times (30 × 30) more energy than a 5.0 magnitude of earthquake. An 8.6 magnitude earthquake releases the same amount of energy as 10,000 atomic bombs like those used in World War II.

This is so because the energy released in an earthquake, and thus its magnitude, is proportional to the area of the fault that ruptures and the stress drop. Therefore, the longer the length and the wider the width of the faulted area, the larger the resulting magnitude. The topmost, brittle part of the Earth's crust, and the cool slabs of the tectonic plates that are descending down into the hot mantle, are the only parts of our planet which can store elastic energy and release it in fault ruptures. Rocks hotter than about 300 degrees Celsius flow in response to stress; they do not rupture in earthquakes. The maximum observed lengths of ruptures and mapped faults (which may break in a single rupture) are approximately 1000 km. Examples are the earthquakes in Chile, 1960; Alaska, 1957; Sumatra, 2004, all in subduction zones. The longest earthquake ruptures on strike-slip faults, like the San Andreas Fault (1857, 1906), the North Anatolian Fault in Turkey (1939) and the Denali Fault in Alaska (2002), are about half to one third as long as the lengths along subducting plate margins, and those along normal faults are even shorter.

The most important parameter controlling the maximum earthquake magnitude on a fault is however not the maximum available length, but the available width because the latter varies by a factor of 20. Along converging plate margins, the dip angle of the rupture plane is very shallow, typically about 10 degrees. Thus the width of the plane within the top brittle crust of the Earth can become 50 to 100 km (Japan, 2011; Alaska, 1964), making the most powerful earthquakes possible.

Aerial photo of the San Andreas Fault in the Carrizo Plain, northwest of Los Angeles

Strike-slip faults tend to be oriented near vertically, resulting in an approximate width of 10 km within the brittle crust, thus earthquakes with magnitudes much larger than 8 are not possible. Maximum magnitudes along many normal faults are even more limited because many of them are located along spreading centers, as in Iceland, where the thickness of the brittle layer is only about 6 km.

In addition, there exists a hierarchy of stress level in the three fault types. Thrust faults are generated by the highest, strike slip by intermediate, and normal faults by the lowest stress levels. This can easily be understood by considering the direction of the greatest principal stress, the direction of the force that 'pushes' the rock mass during the faulting. In the case of normal faults, the rock mass is pushed down in a vertical direction, thus the pushing force (greatest principal stress) equals the weight of the rock mass itself. In the case of thrusting, the rock mass 'escapes' in the direction of the least principal stress, namely upward, lifting the rock mass up, thus the overburden equals the least principal stress. Strike-slip faulting is intermediate between the other two types described above. This difference in stress regime in the three faulting environments can contribute to differences in stress drop during faulting, which contributes to differences in the radiated energy, regardless of fault dimensions.

Earthquakes Away from Plate Boundaries

Where plate boundaries occur within the continental lithosphere, deformation is spread out over a much larger area than the plate boundary itself. In the case of the San Andreas fault continental transform, many earthquakes occur away from the plate boundary and are related to strains developed within the broader zone of deformation caused by major irregularities in the fault trace (e.g., the "Big bend" region). The Northridge earthquake was associated with movement on a blind thrust within such a zone. Another example is the strongly oblique convergent plate boundary between the Arabian and Eurasian plates where it runs through the northwestern part of the Zagros Mountains. The deformation associated with this plate boundary is partitioned into nearly pure thrust sense movements perpendicular to the boundary over a wide zone to the southwest and nearly pure strike-slip motion along the Main Recent Fault close to the actual plate boundary itself. This is demonstrated by earthquake focal mechanisms.

All tectonic plates have internal stress fields caused by their interactions with neighboring plates and sedimentary loading or unloading (e.g. deglaciation). These stresses may be sufficient to cause failure along existing fault planes, giving rise to intraplate earthquakes.

Shallow-focus and Deep-focus Earthquakes

Collapsed Gran Hotel building in the San Salvador metropolis, after the shallow 1986 San Salvador earthquake.

The majority of tectonic earthquakes originate at the ring of fire in depths not exceeding tens of kilometers. Earthquakes occurring at a depth of less than 70 km are classified as 'shallow-focus' earthquakes, while those with a focal-depth between 70 and 300 km are commonly termed 'mid-focus' or 'intermediate-depth' earthquakes. In subduction zones, where older and colder oceanic crust descends beneath another tectonic plate, Deep-focus earthquakes may occur at much greater depths (ranging from 300 up to 700 kilometers). These seismically active areas of subduction are known as Wadati–Benioff zones. Deep-focus earthquakes occur at a depth where the subducted lithosphere should no longer be brittle, due to the high temperature and pressure. A possible mechanism for the generation of deep-focus earthquakes is faulting caused by olivine undergoing a phase transition into a spinel structure.

Earthquakes and Volcanic Activity

Earthquakes often occur in volcanic regions and are caused there, both by tectonic faults and the movement of magma in volcanoes. Such earthquakes can serve as an early warning of volcanic eruptions, as during the 1980 eruption of Mount St. Helens. Earthquake swarms can serve as markers for the location of the flowing magma throughout the volcanoes. These swarms can be recorded by seismometers and tiltmeters (a device that measures ground slope) and used as sensors to predict imminent or upcoming eruptions.

Rupture Dynamics

A tectonic earthquake begins by an initial rupture at a point on the fault surface, a process known as nucleation. The scale of the nucleation zone is uncertain, with some evidence, such as the rupture dimensions of the smallest earthquakes, suggesting that it is smaller than 100 m while other evidence, such as a slow component revealed by low-frequency spectra of some earthquakes, suggest that it is larger. The possibility that the nucleation involves some sort of preparation process is supported by the observation that about 40% of earthquakes are preceded by foreshocks. Once the rupture has initiated, it begins to propagate along the fault surface. The mechanics of this process are poorly understood, partly because it is difficult to recreate the high sliding velocities in a laboratory. Also the effects of strong ground motion make it very difficult to record information close to a nucleation zone.

Rupture propagation is generally modeled using a fracture mechanics approach, likening the rupture to a propagating mixed mode shear crack. The rupture velocity is a function of the fracture energy in the volume around the crack tip, increasing with decreasing fracture energy. The velocity of rupture propagation is orders of magnitude faster than the displacement velocity across the fault. Earthquake ruptures typically propagate at velocities that are in the range 70–90% of the S-wave velocity, and this is independent of earthquake size. A small subset of earthquake ruptures appear to have propagated at speeds greater than the S-wave velocity. These supershear earthquakes have all been observed during large strike-slip events. The unusually wide zone of coseismic damage caused by the 2001 Kunlun earthquake has been attributed to the effects of the sonic boom developed in such earthquakes. Some earthquake ruptures travel at unusually low velocities and are referred to as slow earthquakes. A particularly dangerous form of slow earthquake is the tsunami earthquake, observed where the relatively low felt intensities, caused by the slow propagation speed of some great earthquakes, fail to alert the population of the neighboring coast, as in the 1896 Sanriku earthquake.

Tidal Forces

Tides may induce some seismicity.

Earthquake Clusters

Most earthquakes form part of a sequence, related to each other in terms of location and time. Most earthquake clusters consist of small tremors that cause little to no damage, but there is a theory that earthquakes can recur in a regular pattern.

Aftershocks

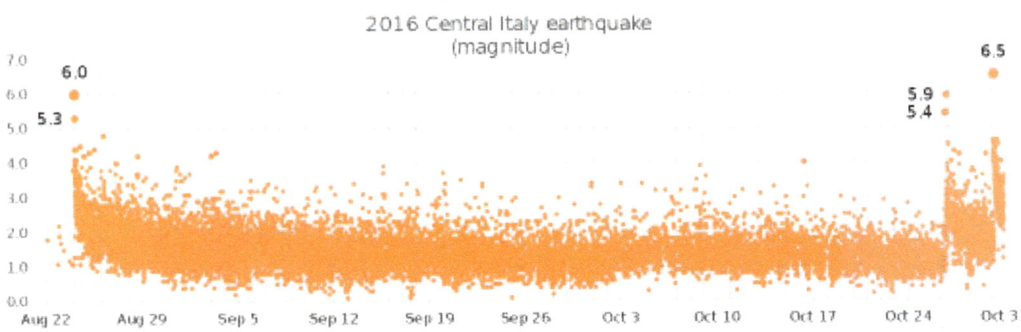

Magnitude of the Central Italy earthquakes of August and October 2016 and the aftershocks (which continued to occur after the period shown here).

An aftershock is an earthquake that occurs after a previous earthquake, the mainshock. An aftershock is in the same region of the main shock but always of a smaller magnitude. If an aftershock is larger than the main shock, the aftershock is redesignated as the main shock and the original main shock is redesignated as a foreshock. Aftershocks are formed as the crust around the displaced fault plane adjusts to the effects of the main shock.

Earthquake Swarms

Earthquake swarms are sequences of earthquakes striking in a specific area within a short period of time. They are different from earthquakes followed by a series of aftershocks by the fact that no single earthquake in the sequence is obviously the main shock, therefore none have notable higher magnitudes than the other. An example of an earthquake swarm is the 2004 activity at Yellowstone National Park. In August 2012, a swarm of earthquakes shook Southern California's Imperial Valley, showing the most recorded activity in the area since the 1970s.

Sometimes a series of earthquakes occur in what has been called an *earthquake storm*, where the earthquakes strike a fault in clusters, each triggered by the shaking or stress redistribution of the previous earthquakes. Similar to aftershocks but on adjacent segments of fault, these storms occur over the course of years, and with some of the later earthquakes as damaging as the early ones. Such a pattern was observed in the sequence of about a dozen earthquakes that struck the North Anatolian Fault in Turkey in the 20th century and has been inferred for older anomalous clusters of large earthquakes in the Middle East.

Size and Frequency of Occurrence

It is estimated that around 500,000 earthquakes occur each year, detectable with current instrumentation. About 100,000 of these can be felt. Minor earthquakes occur nearly constantly around the world in places like California and Alaska in the U.S., as well as in El Salvador, Mexico, Guatemala, Chile, Peru, Indonesia, Iran, Pakistan, the Azores in Portugal, Turkey, New Zealand, Greece, Italy, India, Nepal and Japan, but earthquakes can occur almost anywhere, including Downstate New York, England, and Australia. Larger earthquakes occur less frequently, the relationship being exponential; for example, roughly ten times as many earthquakes larger than magnitude 4 occur in a particular time period than earthquakes larger than magnitude 5. In the (low seismicity) United Kingdom, for example, it has been calculated that the average recurrences are: an earthquake of 3.7–4.6 every year, an earthquake of 4.7–5.5 every 10 years, and an earthquake of 5.6 or larger every 100 years. This is an example of the Gutenberg–Richter law.

The number of seismic stations has increased from about 350 in 1931 to many thousands today. As a result, many more earthquakes are reported than in the past, but this is because of the vast improvement in instrumentation, rather than an increase in the number of earthquakes. The United States Geological Survey estimates that, since 1900, there have been an average of 18 major earthquakes (magnitude 7.0–7.9) and one great earthquake (magnitude 8.0 or greater) per year, and that this average has been relatively stable. In recent years, the number of major earthquakes per year has decreased, though this is probably a statistical fluctuation rather than a systematic trend. More detailed statistics on the size and frequency of earthquakes is available from the United States Geological Survey (USGS). A recent increase in the number of major earthquakes has been noted, which could be explained by a cyclical pattern of periods of intense tectonic activity, interspersed with longer periods of low-intensity. However, accurate recordings of earthquakes only began in the early 1900s, so it is too early to categorically state that this is the case.

The Messina earthquake and tsunami took as many as 200,000 lives on December 28, 1908 in Sicily and Calabria.

Most of the world's earthquakes (90%, and 81% of the largest) take place in the 40,000 km long, horseshoe-shaped zone called the circum-Pacific seismic belt, known as the Pacific Ring of Fire, which for the most part bounds the Pacific Plate. Massive earthquakes tend to occur along other plate boundaries, too, such as along the Himalayan Mountains.

With the rapid growth of mega-cities such as Mexico City, Tokyo and Tehran, in areas of high seismic risk, some seismologists are warning that a single quake may claim the lives of up to 3 million people.

Induced Seismicity

While most earthquakes are caused by movement of the Earth's tectonic plates, human activity can also produce earthquakes. Four main activities contribute to this phenomenon: storing large amounts of water behind a dam (and possibly building an extremely heavy building), drilling and injecting liquid into wells, and by coal mining and oil drilling. Perhaps the best known example is the 2008 Sichuan earthquake in China's Sichuan Province in May; this tremor resulted in 69,227 fatalities and is the 19th deadliest earthquake of all time. The Zipingpu Dam is believed to have fluctuated the pressure of the fault 1,650 feet (503 m) away; this pressure probably increased the power of the earthquake and accelerated the rate of movement for the fault. The greatest earthquake in Australia's history is also claimed to be induced by humanity, through coal mining. The city of Newcastle was built over a large sector of coal mining areas. The earthquake has been reported to be spawned from a fault that reactivated due to the millions of tonnes of rock removed in the mining process.

Measuring and Locating Earthquakes

Earthquakes can be recorded by seismometers up to great distances, because seismic waves travel through the whole Earth's interior. The absolute magnitude of a quake is conventionally reported by numbers on the moment magnitude scale (formerly Richter scale, magnitude 7 causing serious damage over large areas), whereas the felt magnitude is reported using the modified Mercalli intensity scale (intensity II–XII).

Every tremor produces different types of seismic waves, which travel through rock with different velocities:

- Longitudinal P-waves (shock- or pressure waves)
- Transverse S-waves (both body waves)
- Surface waves — (Rayleigh and Love waves)

Propagation velocity of the seismic waves ranges from approx. 3 km/s up to 13 km/s, depending on the density and elasticity of the medium. In the Earth's interior the shock- or P waves travel much faster than the S waves (approx. relation 1.7 : 1). The differences in travel time from the epicenter to the observatory are a measure of the distance and can be used to image both sources of quakes and structures within the Earth. Also the depth of the hypocenter can be computed roughly.

In solid rock P-waves travel at about 6 to 7 km per second; the velocity increases within the deep mantle to ~13 km/s. The velocity of S-waves ranges from 2–3 km/s in light sediments and 4–5 km/s in the Earth's crust up to 7 km/s in the deep mantle. As a consequence, the first waves of a distant earthquake arrive at an observatory via the Earth's mantle.

On average, the kilometer distance to the earthquake is the number of seconds between the P and S wave times 8. Slight deviations are caused by inhomogeneities of subsurface structure. By such analyses of seismograms the Earth's core was located in 1913 by Beno Gutenberg.

Earthquakes are not only categorized by their magnitude but also by the place where they occur. The world is divided into 754 Flinn–Engdahl regions (F-E regions), which are based on political and geographical boundaries as well as seismic activity. More active zones are divided into smaller F-E regions whereas less active zones belong to larger F-E regions.

Standard reporting of earthquakes includes its magnitude, date and time of occurrence, geographic coordinates of its epicenter, depth of the epicenter, geographical region, distances to population centers, location uncertainty, a number of parameters that are included in USGS earthquake reports (number of stations reporting, number of observations, etc.), and a unique event ID.

Although relatively slow seismic waves have traditionally been used to detect earthquakes, scientists realized in 2016 that gravitational measurements could provide instantaneous detection of earthquakes, and confirmed this by analyzing gravitational records associated with the 2011 Tohoku-Oki ("Fukushima") earthquake.

Effects of Earthquakes

The effects of earthquakes include, but are not limited to, the following:

1755 copper engraving depicting Lisbon in ruins and in flames after the 1755 Lisbon earthquake, which killed an estimated 60,000 people. A tsunami overwhelms the ships in the harbor.

Shaking and Ground Rupture

Damaged buildings in Port-au-Prince, Haiti, January 2010.

Shaking and ground rupture are the main effects created by earthquakes, principally resulting in more or less severe damage to buildings and other rigid structures. The severity of the local effects depends on the complex combination of the earthquake magnitude, the distance from the epicenter, and the local geological and geomorphological conditions, which may amplify or reduce wave propagation. The ground-shaking is measured by ground acceleration.

Specific local geological, geomorphological, and geostructural features can induce high levels of shaking on the ground surface even from low-intensity earthquakes. This effect is called site or local amplification. It is principally due to the transfer of the seismic motion from hard deep soils to soft superficial soils and to effects of seismic energy focalization owing to typical geometrical setting of the deposits.

Ground rupture is a visible breaking and displacement of the Earth's surface along the trace of the fault, which may be of the order of several meters in the case of major earthquakes. Ground rupture is a major risk for large engineering structures such as dams, bridges and nuclear power stations and requires careful mapping of existing faults to identify any which are likely to break the ground surface within the life of the structure.

Landslides and Avalanches

Earthquakes, along with severe storms, volcanic activity, coastal wave attack, and wildfires, can produce slope instability leading to landslides, a major geological hazard. Landslide danger may persist while emergency personnel are attempting rescue.

Fires

Fires of the 1906 San Francisco earthquake

Earthquakes can cause fires by damaging electrical power or gas lines. In the event of water mains rupturing and a loss of pressure, it may also become difficult to stop the spread of a fire once it has started. For example, more deaths in the 1906 San Francisco earthquake were caused by fire than by the earthquake itself.

Soil Liquefaction

Soil liquefaction occurs when, because of the shaking, water-saturated granular material (such as sand) temporarily loses its strength and transforms from a solid to a liquid. Soil liquefaction may cause rigid structures, like buildings and bridges, to tilt or sink into the liquefied deposits. For example, in the 1964 Alaska earthquake, soil liquefaction caused many buildings to sink into the ground, eventually collapsing upon themselves.

Tsunami

Tsunamis are long-wavelength, long-period sea waves produced by the sudden or abrupt movement of large volumes of water. In the open ocean the distance between wave crests can surpass 100 kilometers (62 mi), and the wave periods can vary from five minutes to one hour. Such tsunamis travel 600-800 kilometers per hour (373–497 miles per hour), depending on water depth. Large waves produced by an earthquake or a submarine landslide can overrun nearby coastal areas in a matter of minutes. Tsunamis can also travel thousands of kilometers across open ocean and wreak destruction on far shores hours after the earthquake that generated them.

Ordinarily, subduction earthquakes under magnitude 7.5 on the Richter scale do not cause tsunamis, although some instances of this have been recorded. Most destructive tsunamis are caused by earthquakes of magnitude 7.5 or more.

The tsunami of the 2004 Indian Ocean earthquake

Floods

A flood is an overflow of any amount of water that reaches land. Floods occur usually when the volume of water within a body of water, such as a river or lake, exceeds the total capacity of the formation, and as a result some of the water flows or sits outside of the normal perimeter of the body. However, floods may be secondary effects of earthquakes, if dams are damaged. Earthquakes may cause landslips to dam rivers, which collapse and cause floods.

The terrain below the Sarez Lake in Tajikistan is in danger of catastrophic flood if the landslide dam formed by the earthquake, known as the Usoi Dam, were to fail during a future earthquake. Impact projections suggest the flood could affect roughly 5 million people.

Human Impacts

Ruins of the Għajn Ħadid Tower, which collapsed in an earthquake in 1856

An earthquake may cause injury and loss of life, road and bridge damage, general property damage, and collapse or destabilization (potentially leading to future collapse) of buildings. The aftermath may bring disease, lack of basic necessities, mental consequences such as panic attacks, depression to survivors, and higher insurance premiums.

Major Earthquakes

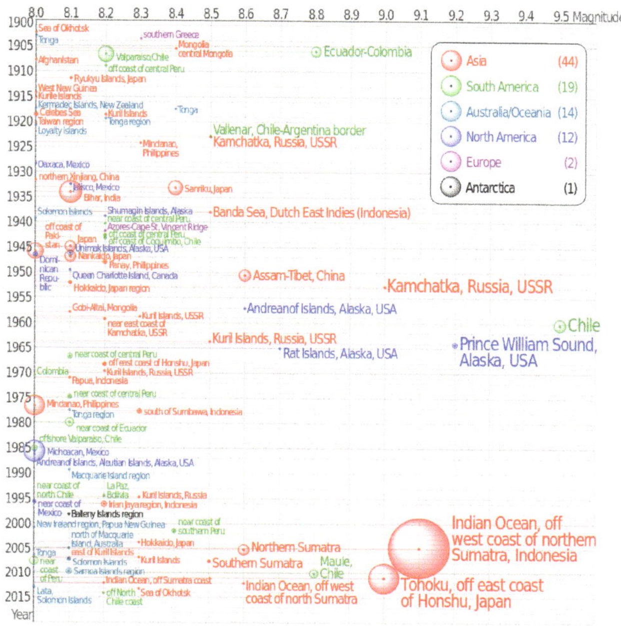

Earthquakes of magnitude 8.0 and greater since 1900. The apparent 3D volumes of the bubbles are linearly proportional to their respective fatalities.

One of the most devastating earthquakes in recorded history was the 1556 Shaanxi earthquake, which occurred on 23 January 1556 in Shaanxi province, China. More than 830,000 people died. Most houses in the area were yaodongs—dwellings carved out of loess hillsides—and many victims were killed when these structures collapsed. The 1976 Tangshan earthquake, which killed between 240,000 and 655,000 people, was the deadliest of the 20th century.

The 1960 Chilean earthquake is the largest earthquake that has been measured on a seismograph, reaching 9.5 magnitude on 22 May 1960. Its epicenter was near Cañete, Chile. The energy released was approximately twice that of the next most powerful earthquake, the Good Friday earthquake (March 27, 1964) which was centered in Prince William Sound, Alaska. The ten largest recorded earthquakes have all been megathrust earthquakes; however, of these ten, only the 2004 Indian Ocean earthquake is simultaneously one of the deadliest earthquakes in history.

Earthquakes that caused the greatest loss of life, while powerful, were deadly because of their proximity to either heavily populated areas or the ocean, where earthquakes often create tsunamis that can devastate communities thousands of kilometers away. Regions most at risk for great loss of life include those where earthquakes are relatively rare but powerful, and poor regions with lax, unenforced, or nonexistent seismic building codes.

Prediction

Many methods have been developed for predicting the time and place in which earthquakes will occur. Despite considerable research efforts by seismologists, scientifically reproducible predictions cannot yet be made to a specific day or month. However, for well-understood faults the probability that a segment may rupture during the next few decades can be estimated.

Earthquake warning systems have been developed that can provide regional notification of an earthquake in progress, but before the ground surface has begun to move, potentially allowing people within the system's range to seek shelter before the earthquake's impact is felt.

Preparedness

The objective of earthquake engineering is to foresee the impact of earthquakes on buildings and other structures and to design such structures to minimize the risk of damage. Existing structures can be modified by seismic retrofitting to improve their resistance to earthquakes. Earthquake insurance can provide building owners with financial protection against losses resulting from earthquakes.

Emergency management strategies can be employed by a government or organization to mitigate risks and prepare for consequences.

Historical Views

Tremblement de terre en Italie. 340 ans avant J.-C. — L. Papirius Cursor consul (d'après Lycosthène).

An image from a 1557 book

From the lifetime of the Greek philosopher Anaxagoras in the 5th century BCE to the 14th century CE, earthquakes were usually attributed to "air (vapors) in the cavities of the Earth." Thales of Miletus, who lived from 625–547 (BCE) was the only documented person who believed that earthquakes were caused by tension between the earth and water. Other theories existed, including the Greek philosopher Anaxamines' (585–526 BCE) beliefs that short incline episodes of dryness and wetness caused seismic activity. The Greek philosopher Democritus (460–371 BCE) blamed water in general for earthquakes. Pliny the Elder called earthquakes "underground thunderstorms."

Recent Studies

In recent studies, geologists claim that global warming is one of the reasons for increased seismic activity. According to these studies melting glaciers and rising sea levels disturb the balance of pressure on Earth's tectonic plates thus causing increase in the frequency and intensity of earthquakes.

Earthquakes in Culture

Mythology and Religion

In Norse mythology, earthquakes were explained as the violent struggling of the god Loki. When Loki, god of mischief and strife, murdered Baldr, god of beauty and light, he was punished by being bound in a cave with a poisonous serpent placed above his head dripping venom. Loki's wife Sigyn stood by him with a bowl to catch the poison, but whenever she had to empty the bowl the poison dripped on Loki's face, forcing him to jerk his head away and thrash against his bonds, which caused the earth to tremble.

In Greek mythology, Poseidon was the cause and god of earthquakes. When he was in a bad mood, he struck the ground with a trident, causing earthquakes and other calamities. He also used earthquakes to punish and inflict fear upon people as revenge.

In Japanese mythology, Namazu is a giant catfish who causes earthquakes. Namazu lives in the mud beneath the earth, and is guarded by the god Kashima who restrains the fish with a stone. When Kashima lets his guard fall, Namazu thrashes about, causing violent earthquakes.

In Popular Culture

In modern popular culture, the portrayal of earthquakes is shaped by the memory of great cities laid waste, such as Kobe in 1995 or San Francisco in 1906. Fictional earthquakes tend to strike suddenly and without warning. For this reason, stories about earthquakes generally begin with the disaster and focus on its immediate aftermath, as in *Short Walk to Daylight* (1972), *The Ragged Edge* (1968) or *Aftershock: Earthquake in New York* (1999). A notable example is Heinrich von Kleist's classic novella, *The Earthquake in Chile*, which describes the destruction of Santiago in 1647. Haruki Murakami's short fiction collection After the Quake depicts the consequences of the Kobe earthquake of 1995.

The most popular single earthquake in fiction is the hypothetical "Big One" expected of California's San Andreas Fault someday, as depicted in the novels *Richter 10* (1996), *Goodbye California* (1977), *2012* (2009) and *San Andreas* (2015) among other works. Jacob M. Appel's widely anthologized short story, *A Comparative Seismology*, features a con artist who convinces an elderly woman that an apocalyptic earthquake is imminent.

Contemporary depictions of earthquakes in film are variable in the manner in which they reflect human psychological reactions to the actual trauma that can be caused to directly afflicted families and their loved ones. Disaster mental health response research emphasizes the need to be aware of the different roles of loss of family and key community members, loss of home and familiar surroundings, loss of essential supplies and services to maintain survival. Particularly for children, the clear availability of caregiving adults who are able to protect, nourish, and clothe them in the aftermath of the earthquake, and to help them make sense of what has befallen them has been shown even more important to their emotional and physical health than the simple giving of provisions. As was observed after other disasters involving destruction and loss of life and their media depictions, recently observed in the 2010 Haiti earthquake, it is also important not to pathologize the reactions to loss and displacement or disruption of governmental administration and services, but rather to validate these reactions, to support constructive problem-solving and reflection as to how one might improve the conditions of those affected.

Avalanche

The toe of an avalanche in Alaska's Kenai Fjords.

A powder snow avalanche in the Himalayas near Mount Everest.

An avalanche (also called a snowslide or snowslip) is a rapid flow of snow down a sloping surface. Avalanches are typically triggered in a starting zone from a mechanical failure in the snowpack (slab avalanche) when the forces on the snow exceed its strength but sometimes only with gradually widening (loose snow avalanche). After initiation, avalanches usually accelerate rapidly and grow in mass and volume as they entrain more snow. If the avalanche moves fast enough some of the snow may mix with the air forming a powder snow avalanche, which is a type of gravity current.

Slides of rocks or debris, behaving in a similar way to snow, are also referred to as avalanches

The load on the snowpack may be only due to gravity, in which case failure may result either from weakening in the snowpack or increased load due to precipitation. Avalanches that occur in this way are known as spontaneous avalanches. Avalanches can also be triggered by other loads such as skiers, snowmobilers, animals or explosives. Seismic activity may also trigger the failure in the snowpack and avalanches. A popular myth is that avalanches can be triggered by loud noise or shouting, but the pressure from sound is orders of magnitude too small to trigger an avalanche.

Although primarily composed of flowing snow and air, large avalanches have the capability to entrain ice, rocks, trees, and other material on the slope, and are distinct from mudslides, rock slides, and serac collapses on an icefall. Avalanches are not rare or random events and are endemic to any mountain range that accumulates a standing snowpack. Avalanches are most common during winter or spring but glacier movements may cause ice and snow avalanches at any time of year. In mountainous terrain, avalanches are among the most serious objective natural hazards to life and property, with their destructive capability resulting from their potential to carry enormous masses of snow at high speeds.

There is no universally accepted classification of avalanches—different classifications are useful for different purposes. Avalanches can be described by their size, their destructive potential, their initiation mechanism, their composition and their dynamics.

Formation and Type

A crown fracture from a slab avalanche near the Neve Glacier in the North Cascades mountains. Extensive fracture propagation is evident.

Loose snow avalanches (far left) and slab avalanches (near center) near Mount Shuksan in the North Cascades mountains. Fracture propagation is relatively limited.

15cm deep, soft slab avalanche triggered by a snowboarder near Heliotrope Ridge, Mount Baker in March 2010. Multiple crown fracture lines are visible in the top-middle of the image. Note the granular characteristic of the debris in the foreground that results from the slab breaking up during descent.

Most avalanches occur spontaneously during storms under increased load due to snowfall. The second largest cause of natural avalanches is metamorphic changes in the snowpack such as melting due to solar radiation. Other natural causes include rain, earthquakes, rockfall and icefall. Artificial triggers of avalanches include skiers, snowmobiles, and controlled explosive work.

Avalanche initiation can start at a point with only a small amount of snow moving initially; this is typical of wet snow avalanches or avalanches in dry unconsolidated snow. However, if the snow has sintered into a stiff slab overlying a weak layer then fractures can propagate very rapidly, so that a large volume of snow, that may be thousands of cubic meters, can start moving almost simultaneously.

A snowpack will fail when the load exceeds the strength. The load is straightforward; it is the weight of the snow. However, the strength of the snowpack is much more difficult to determine and is extremely heterogenous. It varies in detail with properties of the snow grains, size, density, morphology, temperature, water content; and the properties of the bonds between the grains. These properties may all metamorphose in time according to the local humidity, water vapour flux, temperature and heat flux. The top of the snowpack is also extensively influenced by incoming radiation and the local air flow. One of the aims of avalanche research is to develop and validate computer models that can describe the evolution of the seasonal snowpack over time. A complicating factor is the complex interaction of terrain and weather, which causes significant spatial and temporal variability of the depths, crystal forms, and layering of the seasonal snowpack.

Slab Avalanches

Slab avalanches form frequently in snow that has been deposited, or redeposited by wind. They have the characteristic appearance of a block (slab) of snow cut out from its surroundings by fractures. Elements of slab avalanches include the following: a crown fracture at the top of the start zone, flank fractures on the sides of the start zones, and a fracture at the bottom called the staunchwall. The crown and flank fractures are vertical walls in the snow delineating the snow that was entrained in the avalanche from the snow that remained on the slope. Slabs can vary in thickness from a few centimetres to three metres. Slab avalanches account for around 90% of avalanche-related fatalities in backcountry users.

Powder Snow Avalanches

The largest avalanches form turbulent suspension currents known as powder snow avalanches or mixed avalanches. These consist of a powder cloud, which overlies a dense avalanche. They can form from any type of snow or initiation mechanism, but usually occur with fresh dry powder. They can exceed speeds of 300 kilometres per hour (190 mph), and masses of 10,000,000 tonnes (9,800,000 long tons; 11,000,000 short tons); their flows can travel long distances along flat valley bottoms and even uphill for short distances.

Wet Snow Avalanches

In contrast to powder snow avalanches, wet snow avalanches are a low velocity suspension of snow and water, with the flow confined to the track surface (McClung, first edition 1999, page 108). The low speed of travel is due to the friction between the sliding surface of the track and the water saturated flow. Despite the low speed of travel (~10–40 km/h), wet snow avalanches are capable of generating powerful destructive forces, due to the large mass, and density. The body of the flow of a wet snow avalanche can plough through soft snow, and can scour boulders, earth, trees, and other vegetation; leaving exposed, and often scored, ground in the avalanche track. Wet snow avalanches can be initiated from either loose snow releases, or slab releases, and only occur in snow packs that are water saturated and isothermally equilibrated to the melting point of water. The isothermal characteristic of wet snow avalanches has led to the secondary term of isothermal slides found in the literature (for example in Daffern, 1999, page 93). At temperate latitudes wet snow avalanches are frequently associated with climatic avalanche cycles at the end of the winter season, when there is significant daytime warming.

Avalanche Pathway

As an avalanche moves down a slope it follows a certain pathway that is dependent on the slope's degree of steepness and the volume of snow/ice involved in the mass movement. The origin of an avalanche is called the Starting Point and typically occurs on a 30-45 degree slope. The body of the pathway is called the Track of the avalanche and usually occurs on a 20-30 degree slope. When the avalanche loses its momentum and eventually stops it reaches the Runout Zone. This usually occurs when the slope has reached a steepness that is less than 20 degrees. These degrees are not consistently true due to the fact that each avalanche is unique depending on the stability of the snowpack that it was derived from as well as the environmental or human influences that triggered the mass movement.

Terrain, Snowpack, Weather

Doug Fesler and Jill Fredston developed a conceptual model of the three primary elements of avalanches: terrain, weather, and snowpack. Terrain describes the places where avalanches occur, weather describes the meteorological conditions that create the snowpack, and snowpack describes the structural characteristics of snow that make avalanche formation possible.

Terrain

Avalanche formation requires a slope shallow enough for snow to accumulate but steep enough for the snow to accelerate once set in motion by the combination of mechanical failure (of the snowpack) and gravity. The angle of the slope that can hold snow, called the angle of repose, depends on a variety of factors such as crystal form and moisture content. Some forms of drier and colder snow will only stick to shallower slopes, while wet and warm snow can bond to very steep surfaces. In particular, in coastal mountains, such as the Cordillera del Paine region of Patagonia, deep snowpacks collect on vertical and even overhanging rock faces. The slope angle that can allow moving snow to accelerate depends on a variety of factors such as the snow's shear strength (which is itself dependent upon crystal form) and the configuration of layers and inter-layer interfaces.

In steep avalanche-prone terrain, traveling on ridges is generally safer than traversing the slopes.

Avalanche path with 800 metres (2,600 ft) vertical fall in the Glacier Peak Wilderness, Washing-ton State. Avalanche paths in alpine terrain may be poorly defined because of limited vegetation. Below tree line, avalanche paths are often delineated by vegetative trim lines created by past avalanches. The start zone is visible near the top of the image, the track is in the middle of the image and clearly denoted by vegetative trimlines, and the runout zone is shown at the bottom of the image. One possible timeline is as follows: an avalanche forms in the start zone near the ridge, and then descends the track, until coming to rest in the runout zone.

A cornice of snow about to fall. Cracks in the snow are visible in area (1). Area (3) fell soon after this picture was taken, leaving area (2) as the new edge.

The snowpack on slopes with sunny exposures is strongly influenced by sunshine. Diurnal cycles of thawing and refreezing can stabilize the snowpack by promoting settlement. Strong freeze-thaw cycles result in the formation of surface crusts during the night and of unstable surface snow during the day. Slopes in the lee of a ridge or of another wind obstacle accumulate more snow and are more likely to include pockets of deep snow, wind slabs, and cornices, all of which, when disturbed, may result in avalanche formation. Conversely, the snowpack on a windward slope is often much shallower than on a lee slope.

Avalanches and avalanche paths share common elements: a start zone where the avalanche originates, a track along which the avalanche flows, and a runout zone where the avalanche comes to rest. The debris deposit is the accumulated mass of the avalanched snow once it has come to rest in the runout zone. For the image at left, many small avalanches form in this avalanche path every year, but most of these avalanches do not run the full vertical or horizontal length of the path. The frequency with which avalanches form in a given area is known as the return period.

The start zone of an avalanche must be steep enough to allow snow to accelerate once set in motion, additionally convex slopes are less stable than concave slopes, because of the disparity between the tensile strength of snow layers and their compressive strength. The composition and structure of the ground surface beneath the snowpack influences the stability of the snowpack, either being a source of strength or weakness. Avalanches are unlikely to form in very thick forests, but boulders and sparsely distributed vegetation can create weak areas deep within the snowpack through the formation of strong temperature gradients. Full-depth avalanches (avalanches that sweep a slope virtually clean of snow cover) are more common on slopes with smooth ground, such as grass or rock slabs.

Generally speaking, avalanches follow drainages down-slope, frequently sharing drainage features with summertime watersheds. At and below tree line, avalanche paths through drainages are well defined by vegetation boundaries called trim lines, which occur where avalanches have removed trees and prevented regrowth of large vegetation. Engineered drainages, such as the avalanche dam on Mount Stephen in Kicking Horse Pass, have been constructed to protect people and property by redirecting the flow of avalanches. Deep debris deposits from avalanches will collect in catchments at the terminus of a run out, such as gullies and river beds.

Slopes flatter than 25 degrees or steeper than 60 degrees typically have a lower incidence of avalanches. Human-triggered avalanches have the greatest incidence when the snow's angle of repose is between 35 and 45 degrees; the critical angle, the angle at which human-triggered avalanches are most frequent, is 38 degrees. When the incidence of human triggered avalanches is normalized by the rates of recreational use, however, hazard increases uniformly with slope angle, and no significant difference in hazard for a given exposure direction can be found. The rule of thumb is: *A slope that is flat enough to hold snow but steep enough to ski has the potential to generate an avalanche, regardless of the angle.*

Snowpack Structure and Characteristics

After surface hoarfrost becomes buried by later snowfall, the buried hoar layer can be a weak layer upon which upper layers can slide.

The snowpack is composed of ground-parallel layers that accumulate over the winter. Each layer contains ice grains that are representative of the distinct meteorological conditions during which the snow formed and was deposited. Once deposited, a snow layer continues to evolve under the influence of the meteorological conditions that prevail after deposition.

For an avalanche to occur, it is necessary that a snowpack have a weak layer (or instability) below a slab of cohesive snow. In practice the formal mechanical and structural factors related to snowpack instability are not directly observable outside of laboratories, thus the more easily observed properties of the snow layers (e.g. penetration resistance, grain size, grain type, temperature) are used as index measurements of the mechanical properties of the snow (e.g. tensile strength, friction coefficients, shear strength, and ductile strength). This results in two principal sources of uncertainty in determining snowpack stability based on snow structure: First, both the factors influencing snow stability and the specific characteristics of the snowpack vary widely within small areas and time scales, resulting in significant difficulty extrapolating point observations of snow layers across different scales of space and time. Second, the relationship between readily observable snowpack characteristics and the snowpack's critical mechanical properties has not been completely developed.

While the deterministic relationship between snowpack characteristics and snowpack stability is still a matter of ongoing scientific study, there is a growing empirical understanding of the snow composition and deposition characteristics that influence the likelihood of an avalanche. Observation and experience has shown that newly fallen snow requires time to bond with the snow layers beneath it, especially if the new snow falls during very cold and dry conditions. If ambient air temperatures are cold enough, shallow snow above or around boulders, plants, and other discontinuities in the slope, weakens from rapid crystal growth that occurs in the presence of a critical temperature gradient. Large, angular snow crystals are indicators of weak snow, because such crystals have fewer bonds per unit volume than small, rounded crystals that pack tightly together. Consolidated snow is less likely to slough than loose powdery layers or wet isothermal snow; however, consolidated snow is a necessary condition for the occurrence of slab avalanches, and persistent instabilities within the snowpack can hide below well-consolidated surface layers. Uncertainty associated with the empirical understanding of the factors influencing snow stability leads most professional avalanche workers to recommend conservative use of avalanche terrain relative to current snowpack instability.

Weather

After digging a snow pit, it is possible to evaluate the snowpack for unstable layers. In this picture, snow from a weak layer has been easily scraped away by hand, leaving a horizontal line in the wall of the pit.

Avalanches can only occur in a standing snowpack. Typically winter seasons at high latitudes, high altitudes, or both, have weather that is sufficiently unsettled, and cold enough for precipitated snow to accumulate into a seasonal snowpack. Continentality, through its potentiating influence on the meteorological extremes experienced by snowpacks, is an important factor in the evolution of instabilities, and consequential occurrence of avalanches. Conversely, proximity to coastal environments moderates the meteorological extremes experienced by snowpacks, and results in a faster stabilization of the snowpack after storm cycles. The evolution of the snowpack is critically sensitive to small variations within the narrow range of meteorological conditions that allow for the accumulation of snow into a snowpack. Among the critical factors controlling snowpack evolution are: heating by the sun, radiational cooling, vertical temperature gradients in standing snow, snowfall amounts, and snow types. Generally, mild winter weather will promote the settlement and stabilization of the snowpack; and conversely very cold, windy, or hot weather will weaken the snowpack.

At temperatures close to the freezing point of water, or during times of moderate solar radiation, a gentle freeze-thaw cycle will take place. The melting and refreezing of water in the snow strengthens the snowpack during the freezing phase and weakens it during the thawing phase. A rapid rise in temperature, to a point significantly above the freezing point of water, may cause avalanche formation at any time of year.

Persistent cold temperatures can either prevent new snow from stabilizing or destabilize the existing snowpack. Cold air temperatures on the snow surface produce a temperature gradient in the snow, because the ground temperature at the base of the snowpack is usually around °C, and the ambient air temperature can be much colder. When a temperature gradient greater than 10 °C change per vertical meter of snow is sustained for more than a day, angular crystals called depth hoar or facets begin forming in the snowpack because of rapid moisture transport along the temperature gradient. These angular crystals, which bond poorly to one another and the surrounding snow, often become a persistent weakness in the snowpack. When a slab lying on top of a persistent weakness is loaded by a force greater than the strength of the slab and persistent weak layer, the persistent weak layer can fail and generate an avalanche.

Any wind stronger than a light breeze can contribute to a rapid accumulation of snow on sheltered slopes downwind. Wind slab forms quickly and, if present, weaker snow below the slab may not have time to adjust to the new load. Even on a clear day, wind can quickly load a slope with snow by blowing snow from one place to another. Top-loading occurs when wind deposits snow from the top of a slope; cross-loading occurs when wind deposits snow parallel to the slope. When a wind blows over the top of a mountain, the leeward, or downwind, side of the mountain experiences top-loading, from the top to the bottom of that lee slope. When the wind blows across a ridge that leads up the mountain, the leeward side of the ridge is subject to cross-loading. Cross-loaded wind-slabs are usually difficult to identify visually.

Snowstorms and rainstorms are important contributors to avalanche danger. Heavy snowfall will cause instability in the existing snowpack, both because of the additional weight and because the new snow has insufficient time to bond to underlying snow layers. Rain has a similar effect. In the short-term, rain causes instability because, like a heavy snowfall, it imposes an additional load on the snowpack; and, once rainwater seeps down through the snow, it acts as a lubricant, reducing the natural friction between snow layers that holds the snowpack together. Most avalanches happen during or soon after a storm.

Daytime exposure to sunlight will rapidly destabilize the upper layers of the snowpack if the sunlight is strong enough to melt the snow, thereby reducing its hardness. During clear nights, the snowpack can re-freeze when ambient air temperatures fall below freezing, through the process of long-wave radiative cooling, or both. Radiative heat loss occurs when the night air is significantly cooler than the snowpack, and the heat stored in the snow is re-radiated into the atmosphere.

Dynamics

When a slab avalanche forms, the slab disintegrates into increasingly smaller fragments as the snow travels downhill. If the fragments become small enough the outer layer of the avalanche, called a saltation layer, takes on the characteristics of a fluid. When sufficiently fine particles are present they can become airborne and, given a sufficient quantity of airborne snow, this portion of the avalanche can become separated from the bulk of the avalanche and travel a greater distance as a powder snow avalanche. Scientific studies using radar, following the 1999 Galtür avalanche disaster, confirmed the hypothesis that a saltation layer forms between the surface and the airborne components of an avalanche, which can also separate from the bulk of the avalanche.

Driving an avalanche is the component of the avalanche's weight parallel to the slope; as the avalanche progresses any unstable snow in its path will tend to become incorporated, so increasing the overall weight. This force will increase as the steepness of the slope increases, and diminish as the slope flattens. Resisting this are a number of components that are thought to interact with each other: the friction between the avalanche and the surface beneath; friction between the air and snow within the fluid; fluid-dynamic drag at the leading edge of the avalanche; shear resistance between the avalanche and the air through which it is passing, and shear resistance between the fragments within the avalanche itself. An avalanche will continue to accelerate until the resistance exceeds the forward force.

Modeling

Attempts to model avalanche behaviour date from the early 20th century, notably the work of Professor Lagotala in preparation for the 1924 Winter Olympics in Chamonix. His method was developed by A. Voellmy and popularised following the publication in 1955 of his *Ueber die Zerstoerungskraft von Lawinen* (On the Destructive Force of Avalanches).

Voellmy used a simple empirical formula, treating an avalanche as a sliding block of snow moving with a drag force that was proportional to the square of the speed of its flow:

$$\text{Pref} = \frac{1}{2}\rho v^2$$

He and others subsequently derived other formulae that take other factors into account, with the Voellmy-Salm-Gubler and the Perla-Cheng-McClung models becoming most widely used as simple tools to model flowing (as opposed to powder snow) avalanches.

Since the 1990s many more sophisticated models have been developed. In Europe much of the recent work was carried out as part of the SATSIE (Avalanche Studies and Model Validation in Europe) research project supported by the European Commission which produced the leading-edge MN2L model, now in use with the *Service Réstitution Terrains en Montagne* (Mountain Rescue Service) in France, and D2FRAM (Dynamical Two-Flow-Regime Avalanche Model), which was still undergoing validation as of 2007. Other known models are the SAMOS-AT avalanche simulation software and the RAMMS software.

Human Involvement

United States Forest Service avalanche danger advisories.

Snow fences in Switzerland

Avalanche blasting in French ski resort Tignes (3,600 m)

Prevention

Preventative measures are employed in areas where avalanches pose a significant threat to people, such as ski resorts, mountain towns, roads, and railways. There are several ways to prevent avalanches and lessen their power and destruction; active preventative measures reduce the likelihood and size of avalanches by disrupting the structure of the snowpack, while passive measures reinforce and stabilize the snowpack *in situ*. The simplest active measure is repeatedly traveling on a snowpack as snow accumulates; this can be by means of boot-packing, ski-cutting, or machine grooming. Explosives are used extensively to prevent avalanches, by triggering smaller avalanches that break down instabilities in the snowpack, and removing overburden that can result in larger avalanches. Explosive charges are delivered by a number of methods including hand-tossed charges, helicopter-dropped bombs, Gazex concussion lines, and ballistic projectiles launched by air cannons and artillery. Passive preventive systems such as snow fences and light walls can be used to direct the placement of snow. Snow builds up around the fence, especially the side that faces the prevailing winds. Downwind of the fence, snow buildup is lessened. This is caused by the loss of snow at the fence that would have been deposited and the pickup of the snow that is already there by the wind, which was depleted of snow at the fence. When there is a sufficient density of trees, they can greatly reduce the strength of avalanches. They hold snow in place and when there is an avalanche, the impact of the snow against the trees slows it down. Trees can either be planted or they can be conserved, such as in the building of a ski resort, to reduce the strength of avalanches.

Mitigation

In many areas, regular avalanche tracks can be identified and precautions can be taken to minimise damage, such as the prevention of development in these areas. To mitigate the effect of avalanches the construction of artificial barriers can be very effective in reducing avalanche damage. There are several types: One kind of barrier (snow net) uses a net strung between poles that are anchored by guy wires in addition to their foundations. These barriers are similar to those used for rockslides. Another type of barrier is a rigid fence-like structure (snow fence) and may be constructed of steel, wood or pre-stressed concrete. They usually have gaps between the beams and are built perpendicular to the slope, with reinforcing beams on the downhill side. Rigid barriers are often considered unsightly, especially when many rows must be built. They are also expensive and vulnerable to damage from falling rocks in the warmer months. In addition to industrially manufactured barriers, landscaped barriers, called avalanche dams stop or deflect avalanches with their weight and strength. These barriers are made out of concrete, rocks or earth. They are usually placed right above the structure, road or railway that they are trying to protect, although they can also be used to channel avalanches into other barriers. Occasionally, earth mounds are placed in the avalanche's path to slow it down. Finally, along transportation corridors, large shelters, called snow sheds, can be built directly in the slide path of an avalanche to protect traffic from avalanches.

Survival, Rescue, and Recovery

Avalanche accidents are broadly differentiated into 2 categories: accidents in recreational settings, and accidents in residential, industrial, and transportation settings. This distinction is motivated by the observed difference in the causes of avalanche accidents in the two settings. In the recreational setting most accidents are caused by the people involved in the avalanche. In a 1996 study, Jamieson et al. (pages 7–20) found that 83% of all avalanches in the recreational setting were caused by those who were involved in the accident. In contrast, all of the accidents in the residential, industrial, and transportation settings were due to spontaneous natural avalanches. Because of the difference in the causes of avalanche accidents, and the activities pursued in the two settings, avalanche and disaster management professionals have developed two related preparedness, rescue, and recovery strategies for each of the settings.

Notable Avalanches

Two avalanches occurred in March 1910 in the Cascade and Selkirk Mountain ranges; On March 1 the Wellington avalanche killed 96 in Washington State, United States. Three days later 62 railroad workers were killed in the Rogers Pass avalanche in British Columbia, Canada.

During World War I, an estimated 40,000 to 80,000 soldiers died as a result of avalanches during the mountain campaign in the Alps at the Austrian-Italian front, many of which were caused by artillery fire. Some 10,000 men, from both sides, lost their lives in avalanches in December 1916.

In the northern hemisphere winter of 1950–1951 approximately 649 avalanches were recorded in a three-month period throughout the Alps in Austria, France, Switzerland, Italy and Germany. This series of avalanches killed around 265 people and was termed the Winter of Terror.

A mountain climbing camp on Lenin Peak, in what is now Kyrgyzstan, was wiped out in 1990 when an earthquake triggered a large avalanche that overran the camp. Forty-three climbers were killed.

In 1993, the Bayburt Üzengili avalanche killed 60 individuals in Üzengili in the province of Bayburt, Turkey.

A large avalanche in Montroc, France, in 1999, 300,000 cubic metres of snow slid on a 30° slope, achieving a speed in the region of 100 km/h (62 mph). It killed 12 people in their chalets under 100,000 tons of snow, 5 meters (16 feet) deep. The mayor of Chamonix was convicted of second-degree murder for not evacuating the area, but received a suspended sentence.

The small Austrian village of Galtür was hit by the Galtür avalanche in 1999. The village was thought to be in a safe zone but the avalanche was exceptionally large and flowed into the village. Thirty-one people died.

On December 1st, 2000, the Glory Bowl Avalanche formed on Mt. Glory which is located within the Teton Mountain Range in Wyoming, United States. Joel Roof was snowboarding recreationally in this backcountry, bowl-shaped run and triggered the avalanche. He was carried nearly 2,000 feet to the base of the mountain and was not successfully rescued.

Classification

European Avalanche Risk Table

In Europe, the avalanche risk is widely rated on the following scale, which was adopted in April 1993 to replace the earlier non-standard national schemes. Descriptions were last updated in May 2003 to enhance uniformity.

In France, most avalanche deaths occur at risk levels 3 and 4. In Switzerland most occur at levels 2 and 3. It is thought that this may be due to national differences of interpretation when assessing the risks.

Risk Level	Snow Stability	Flag	Avalanche Risk
1 - Low	Snow is generally very stable.		Avalanches are unlikely except when heavy loads are applied on a very few extreme steep slopes. Any spontaneous avalanches will be minor sloughs. In general, safe conditions.
2 - Moderate	On some steep slopes the snow is only moderately stable. Elsewhere it is very stable.		Avalanches may be triggered when heavy loads are applied, especially on a few generally identified steep slopes. Large spontaneous avalanches are not expected.
3 - Considerable	On many steep slopes the snow is only moderately or weakly stable.		Avalanches may be triggered on many slopes even if only light loads are applied. On some slopes, medium or even fairly large spontaneous avalanches may occur.
4 - High	On most steep slopes the snow is not very stable.		Avalanches are likely to be triggered on many slopes even if only light loads are applied. In some places, many medium or sometimes large spontaneous avalanches are likely.

5 - Very High	The snow is generally unstable.		Even on gentle slopes, many large spontaneous avalanches are likely to occur.

Stability:

- Generally described in more detail in the avalanche bulletin (regarding the altitude, aspect, type of terrain etc.)

Additional Load:

- heavy: two or more skiers or boarders without spacing between them, a single hiker or climber, a grooming machine, avalanche blasting
- light: a single skier or snowboarder smoothly linking turns and without falling, a group of skiers or snowboarders with a minimum 10 m gap between each person, a single person on snowshoes

Gradient:

- gentle slopes: with an incline below about 30°
- steep slopes: with an incline over 30°
- very steep slopes: with an incline over 35°
- extremely steep slopes: extreme in terms of the incline (over 40°), the terrain profile, proximity of the ridge, smoothness of underlying ground

European Avalanche Size Table

Avalanche Size:

Size	Runout	Potential Damage	Physical Size
1 - Sluff	Small snow slide that cannot bury a person, though there is a danger of falling.	Unlikely, but possible risk of injury or death to people.	length <50 m volume <100 m³
2 - Small	Stops within the slope.	Could bury, injure or kill a person.	length <100 m volume <1,000 m³
3 - Medium	Runs to the bottom of the slope.	Could bury and destroy a car, damage a truck, destroy small buildings or break trees.	length <1,000 m volume <10,000 m³
4 - Large	Runs over flat areas (significantly less than 30°) of at least 50 m in length, may reach the valley bottom.	Could bury and destroy large trucks and trains, large buildings and forested areas.	length >1,000 m volume >10,000 m³

North American Avalanche Danger Scale

In the United States and Canada, the following avalanche danger scale is used. Descriptors vary depending on country.

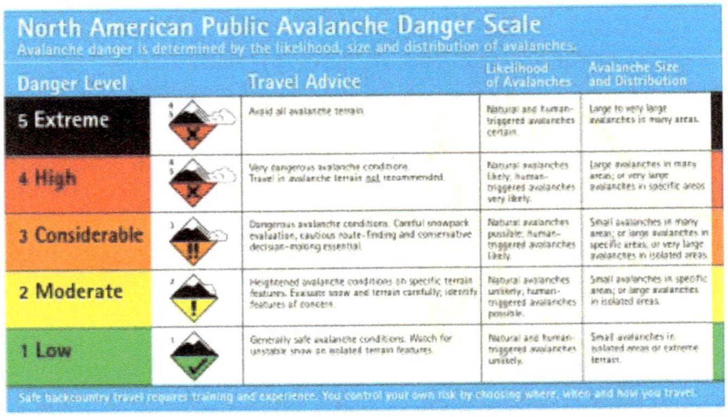

North American Public Avalanche Danger Scale

Avalanche danger is determined by the likelihood, size and distribution of avalanches.

Danger Level	Travel Advice	Likelihood of Avalanches	Avalanche Size and Distribution
5 Extreme	Avoid all avalanche terrain.	Natural and human-triggered avalanches certain.	Large to very large avalanches in many areas.
4 High	Very dangerous avalanche conditions. Travel in avalanche terrain not recommended.	Natural avalanches likely; human-triggered avalanches very likely.	Large avalanches in many areas; or very large avalanches in specific areas.
3 Considerable	Dangerous avalanche conditions. Careful snowpack evaluation, cautious route-finding and conservative decision-making essential.	Natural avalanches possible; human-triggered avalanches likely.	Small avalanches in many areas; or large avalanches in specific areas; or very large avalanches in isolated areas.
2 Moderate	Heightened avalanche conditions on specific terrain features. Evaluate snow and terrain carefully; identify features of concern.	Natural avalanches unlikely; human-triggered avalanches possible.	Small avalanches in specific areas; or large avalanches in isolated areas.
1 Low	Generally safe avalanche conditions. Watch for unstable snow on isolated terrain features.	Natural and human-triggered avalanches unlikely.	Small avalanches in isolated areas or extreme terrain.

Safe backcountry travel requires training and experience. You control your own risk by choosing where, when and how you travel.

Danger Scale - English

Canadian Classification for Avalanche Size

The Canadian classification for avalanche size is based upon the consequences of the avalanche. Half sizes are commonly used.

Size	Destructive Potential
1	Relatively harmless to people.
2	Could bury, injure or kill a person.
3	Could bury and destroy a car, damage a truck, destroy a small building or break a few trees.
4	Could destroy a railway car, large truck, several buildings or a forest area up to 4 hectares.
5	Largest snow avalanche known. Could destroy a village or a forest of 40 hectares.

United States Classification for Avalanche Size

Size	Destructive Potential
1	Sluff or snow that slides less than 50m (150') of slope distance.
2	Small, relative to path.
3	Medium, relative to path.
4	Large, relative to path.
5	Major or maximum, relative to path.

Rutschblock Test

Slab avalanche hazard analysis can be done using the Rutschblock Test. A 2 m wide block of snow is isolated from the rest of the slope and progressively loaded. The result is a rating of slope stability on a seven step scale. (Rutsch means slide in German).

Sinkhole

A sinkhole, also known as a cenote, sink, sink-hole, shakehole, swallet, swallow hole, or doline (the different terms for sinkholes are often used interchangeably), is a depression or hole in the ground caused by some form of collapse of the surface layer. Most are caused by karst processes—for example, the chemical dissolution of carbonate rocks or suffosion processes. Sinkholes vary in size from 1 to 600 m (3.3 to 2,000 ft) both in diameter and depth, and vary in form from soil-lined bowls to bedrock-edged chasms. Sinkholes may form gradually or suddenly, and are found worldwide.

The Red Lake sinkhole in Croatia

Formation

Sinkholes near the Dead Sea, formed when underground salt is dissolved by freshwater intrusion, due to continuing sea-level drop.

Collapse sinkhole in gypsum, near Madrid, central Spain.

Natural Processes

Sinkholes may capture surface drainage from running or standing water, but may also form in high and dry places in specific locations. Sinkholes that capture drainage can hold it in large limestone caves. These caves may drain into tributaries of larger rivers.

The formation of sinkholes involves natural processes of erosion or gradual removal of slightly soluble bedrock (such as limestone) by percolating water, the collapse of a cave roof, or a lowering of the water table. Sinkholes often form through the process of suffosion. For example, groundwater may dissolve the carbonate cement holding the sandstone particles together and then carry away the lax particles, gradually forming a void.

Occasionally a sinkhole may exhibit a visible opening into a cave below. In the case of exceptionally large sinkholes, such as the Minyé sinkhole in Papua New Guinea or Cedar Sink at Mammoth Cave National Park in Kentucky, an underground stream or river may be visible across its bottom flowing from one side to the other.

Sinkholes are common where the rock below the land surface is limestone or other carbonate rock, salt beds, or in other soluble rocks, such as gypsum, that can be dissolved naturally by circulating ground water. Sinkholes also occur in sandstone and quartzite terrains.

As the rock dissolves, spaces and caverns develop underground. These sinkholes can be dramatic, because the surface land usually stays intact until there is not enough support. Then, a sudden collapse of the land surface can occur.

On 2 July 2015, scientists reported that active pits, related to sinkhole collapses and possibly associated with outbursts, were found on the comet 67P/Churyumov-Gerasimenko by the *Rosetta* space probe.

Artificial Processes

Sinkhole formed by rainwater leaking through pavement and carrying soil into a ruptured sewer pipe.

Sinkholes also form from human activity, such as the collapse of abandoned mines and salt cavern storage in salt domes in places like Louisiana, Mississippi and Texas. More commonly, sinkholes occur in urban areas due to water main breaks or sewer collapses when old pipes give way. They can also occur from the overpumping and extraction of groundwater and subsurface fluids.

Sinkholes can also form when natural water-drainage patterns are changed and new water-diversion systems are developed. Some sinkholes form when the land surface is changed, such as when industrial and runoff-storage ponds are created; the substantial weight of the new material can trigger an underground collapse of supporting material, thus causing a sinkhole.

Occurrence

The entire surface water flow of the Alapaha River near Jennings, Florida goes into a sinkhole leading to the Floridan Aquifer groundwater

Sinkholes tend to occur in karst landscapes. Karst landscapes can have up to thousands of sinkholes within a small area, giving the landscape a pock-marked appearance. These sinkholes drain all the water, so there are only subterranean rivers in these areas. Examples of karst landscapes with a plethora of massive sinkholes include Khammouan Mountains (Laos) and Mamo Plateau (Papua New Guinea). The largest known sinkholes formed in sandstone are Sima Humboldt and Sima Martel in Venezuela.

Some sinkholes form in thick layers of homogenous limestone. Their formation is facilitated by high groundwater flow, often caused by high rainfall; such rainfall causes formation of the giant sinkholes in the Nakanaï Mountains, on the New Britain island in Papua New Guinea. On the contact of limestone and insoluble rock below it, powerful underground rivers may form, creating large underground voids.

In such conditions, the largest known sinkholes of the world have formed, like the 662-metre (2,172 ft) deep Xiaozhai Tiankeng (Chongqing, China), giant sótanos in Querétaro and San Luis Potosí states in Mexico and others.

Unusual processes have formed the enormous sinkholes of Sistema Zacatón in Tamaulipas (Mexico), where more than 20 sinkholes and other karst formations have been shaped by volcanically heated, acidic groundwater. This has produced not only the formation of the deepest water-filled sinkhole in the world—Zacatón—but also unique processes of travertine sedimentation in upper parts of sinkholes, leading to sealing of these sinkholes with travertine lids.

The state of Florida in the United States is known for having frequent sinkhole collapses, especially in the central part of the state. The Murge area in southern Italy also has numerous sinkholes. Sinkholes can be formed in retention ponds from large amounts of rain.

Human Uses

Sinkholes have been used for centuries as disposal sites for various forms of waste. A consequence of this is the pollution of groundwater resources, with serious health implications in such areas.

The Maya civilization sometimes used sinkholes in the Yucatán Peninsula (known as cenotes) as places to deposit precious items and human sacrifices.

When sinkholes are very deep or connected to caves, they may offer challenges for experienced cavers or, when water-filled, divers. Some of the most spectacular are the Zacatón cenote in Mexico (the world's deepest water-filled sinkhole), the Boesmansgat sinkhole in South Africa, Sarisariñama tepuy in Venezuela, the Sótano del Barro in Mexico, and in the town of Mount Gambier, South Australia. Sinkholes that form in coral reefs and islands that collapse to enormous depths are known as blue holes and often become popular diving spots.

Local Names

The Great Blue Hole near Ambergris Caye, Belize

Large and visually unusual sinkholes have been well-known to local people since ancient times. Nowadays sinkholes are grouped and named in site-specific or generic names. Some examples of such names are listed below.

- Black holes – This term refers to a group of unique, round, water-filled pits in the Bahamas. These formations seem to be dissolved in carbonate mud from above, by the sea water. The dark color of the water is caused by a layer of phototropic microorganisms concentrated in a dense, purple colored layer at 15 to 20 m (49 to 66 ft) depth; this layer "swallows" the light. Metabolism in the layer of microorganisms causes heating of the water, the only known case in the natural world where microorganisms create significant thermal effects. One of them is the Black Hole of Andros.

- Blue holes – This name was initially given to the deep underwater sinkholes of the Bahamas but is often used for any deep water-filled pits formed in carbonate rocks. The name originates from the deep blue color of water in these sinkholes, which is created by the high clarity of the water and the great depth of the sinkholes; only the deep blue color of the visible spectrum can penetrate such depth and return after reflection.

- Cenotes – This refers to the characteristic water-filled sinkholes in the Yucatán Peninsula,

Belize and some other regions. Many cenotes have formed in limestone deposited in shallow seas created by the Chicxulub meteorite's impact.

- Sótanos – This name is given to several giant pits in several states of Mexico.
- Tiankengs – These are extremely large sinkholes, typically deeper and wider than 250 m (820 ft), with mostly vertical walls, most often created by the collapse of underground caverns. The term means *sky holes* in Chinese; many of this largest type of sinkhole are located in China.
- Tomo – This term is used in New Zealand karst country to describe pot holes.

Piping Pseudokarst

The 2010 Guatemala City sinkhole formed suddenly in May of that year; torrential rains from Tropical Storm Agatha and a bad drainage system were blamed for its creation. It swallowed a three-story building and a house; it measured approximately 20 m (66 ft) wide and 30 m (98 ft) deep. A similar hole had formed nearby in February 2007.

This large vertical hole is not a true sinkhole, as it did not form via the dissolution of limestone, dolomite, marble, or any other water-soluble rock. Guatemala City is not underlain by any carbonate rock; instead, thick deposits of volcanic ash, unwelded ash flow tuffs, and other pyroclastic debris underlie all of Guatemala City. The dissolution of rock did not form the large vertical holes that swallowed up parts of Guatemala City in 2007 and 2010.

The Guatemala City holes are instead an example of "piping pseudokarst", created by the collapse of large cavities that had developed in the weak, crumbly Quaternary volcanic deposits underlying the city. Although weak and crumbly, these volcanic deposits have enough cohesion to allow them to stand in vertical faces and to develop large subterranean voids within them. A process called "soil piping" first created large underground voids, as water from leaking water mains flowed through these volcanic deposits and mechanically washed fine volcanic materials out of them, then progressively eroded and removed coarser materials. Eventually, these underground voids became large enough that their roofs collapsed to create large holes.

Notable Examples

Bimmah or Falling Star Sinkhole in Oman

Some of the largest sinkholes in the world are:

In Africa

- Blue Hole – Dahab, Egypt. A round sinkhole or blue hole, 130 m (430 ft) deep. It includes an archway leading out to the Red Sea at 60 m (200 ft), which has been the site for many freediving and scuba attempts, the latter often fatal.
- Boesmansgat – South African freshwater sinkhole, approximately 290 m (950 ft) deep.
- Lake Kashiba – Zambia. About 3.5 hectares (8.6 acres) in area and about 100 m (330 ft) deep.

In Asia

- Akhayat sinkhole is in Mersin Province, Turkey. Its dimensions are about 150 m (490 f) in diameter with a maximum depth of 70 m (230 ft).
- Bimmah Sinkhole (Hawiyat Najm, the Falling Star Sinkhole, Dibab Sinkhole) – Oman, approximately 30 m (98 ft) deep.
- Dashiwei Tiankeng in Guangxi, China, is 613 m (2,011 ft) deep, with vertical walls. At the bottom is an isolated patch of forest with rare species.
- Dragon Hole the deepest, known, underwater sinkhole in the world.
- Shaanxi tiankeng cluster, in the Daba Mountains of southern Shaanxi, China, covers an area of nearly 5019 square kilometers with the largest sinkhole being 520 meters in diameter and 320 meters deep.
- Teiq Sinkhole (Taiq, Teeq, Tayq) in Oman is one of the largest sinkholes in the world by volume: 90,000,000 m^3 (3.2×10^9 cu ft). Several perennial wadis fall with spectacular waterfalls into this 250 m (820 ft) deep sinkhole.
- Xiaozhai Tiankeng – Chongqing Municipality, China. Double nested sinkhole with vertical walls, 662 m (2,172 ft) deep.

In the Caribbean

- Dean's Blue Hole – Bahamas. The second deepest known sinkhole under the sea, depth 203 m (666 ft). Popular location for world championships of free diving, as well as recreational diving.

In Central America

- Great Blue Hole – Belize. Spectacular, round sinkhole, 124 m (407 ft) deep. Unusual features are tilted stalactites in great depth, which mark the former orientation of limestone layers when this sinkhole was above sea level.
- 2007 Guatemala City sinkhole
- 2010 Guatemala City sinkhole

In Europe

- Hranice Abyss, Moravia, Czech Republic, is the deepest known underwater cave in the world, the lowest confirmed depth (as of 27 September 2016) is 473 m (404 m under the water level).
- Pozzo del Merro, near Rome, Italy. At the bottom of an 80 m (260 ft) conical pit, and approximately 400 m (1,300 ft) deep, it is among the deepest sinkholes in the world.
- Red Lake – Croatia. Approximately 530 m (1,740 ft) deep pit with nearly vertical walls, contains an approximately 280–290 m (920–950 ft) deep lake.
- Vouliagmeni – Greece. The sinkhole of Vouliagmeni is known as "The Devil Well", because it is considered extremely dangerous. Four scuba divers are known to have died in it. Maximum depth of 35.2 m (115 ft) and horizontal penetration of 150 m (490 ft).
- Pouldergaderry – Ireland. This sinkhole is located in the townland of Kilderry South near the Miltown, Co. Kerry at 52°7′57.5″N 9°44′45.4″W / 52.132639°N 9.745944°W / 52.132639; -9.745944. The sinkhole, which is located in an area of karst bedrock, is approximately 80 metres (260 ft) in diameter and 30 metres (98 ft) deep with many mature trees growing on the floor of the hole. At the level of the surrounding ground, the sinkhole covers an area of approximately 1.3 acres. Its presence is indicated on Ordnance Survey maps dating back to 1829.

In North America

Mexico

- Cave of Swallows – San Luis Potosí. 372 m (1,220 ft) deep, round sinkhole with overhanging walls.
- Sima de las Cotorras – Chiapas. 160 m (520 ft) across, 140 m (460 ft) deep, with thousands of green parakeets and ancient rock paintings.
- Sótano de la Lucha – Chiapas. Bigger than Sima de las Cotorras and with lush vegetation on the floor. It can be reached through a cave.
- Sótano del Barro – Querétaro. 410 m (1,350 ft) deep, with nearly vertical walls.
- Zacatón – Tamaulipas. Deepest water-filled sinkhole in world, 339 m (1,112 ft) deep.

United States

- Bayou Corne sinkhole – Assumption Parish, Louisiana. About 25 acres in area and 750 ft (230 m) deep.
- The Blue Hole – Santa Rosa, New Mexico. The surface entrance is only 80 feet (24 m) in diameter, it expands to a diameter of 130 feet (40 m) at the bottom.
- Daisetta Sinkholes – Daisetta, Texas. Several sinkholes have formed, the most recent in 2008 with a maximum diameter of 620 ft (190 m) and maximum depth of 150 ft (46 m).

- Devil's Millhopper – Gainesville, Florida. 120 ft (37 m) deep, 500 ft (150 m) wide. Twelve springs, some more visible than others, feed a pond at the bottom.
- "Golly Hole" (or "December Giant"), Calera, Alabama, appeared December 2, 1972. Approximately 300 ft (91 m) by 325 ft (99 m) and 120 ft (37 m) deep.
- Grassy Cove – Tennessee. 13.6 km² (5.3 sq mi) in area and 42.7 m (140 ft) deep, a National Natural Landmark.
- Gypsum Sinkhole – Utah, in Capitol Reef National Park. Nearly 15 m (49 ft) in diameter and approximately 60 m (200 ft) deep.
- Kingsley Lake – Florida. 8.1 km² (2,000 acres) in area, 27 m (89 ft) deep and almost perfectly round.
- Lake Peigneur – New Iberia, Louisiana. Original depth 11 ft (3.4 m), currently 1,300 ft (400 m) at Diamond Crystal Salt Mine collapse.
- Winter Park Sinkhole, in central Florida, appeared May 8, 1981. It was approximately 350 feet (107 m) wide and 75 feet (23 m) deep. It was notable as one of the largest recent sinkholes to form in the United States. It is now known as Lake Rose.

In Oceania

- Harwood Hole – Abel Tasman National Park, New Zealand, 183 m (600 ft) deep.
- Minyé sinkhole – East New Britain, Papua New Guinea. 510 m (1,670 ft) deep, with vertical walls, crossed by a powerful stream.

In South America

- Sima Humboldt – Venezuela. Largest sinkhole in sandstone, 314 m (1,030 ft) deep, with vertical walls. Unique, isolated forest on bottom.

Limnic Eruption

A limnic eruption, also referred to as a lake overturn, is a rare type of natural disaster in which dissolved carbon dioxide (CO_2) suddenly erupts from deep lake waters, forming a gas cloud that can suffocate wildlife, livestock and humans. Such an eruption may also cause tsunamis in the lake as the rising CO_2 displaces water. Scientists believe earthquakes, volcanic activity, or explosions can be a trigger for such phenomenon. Lakes in which such activity occurs may be known as *limnically active lakes* or *exploding lakes*. Some features of limnically active lakes include:

- CO_2-saturated incoming water
- A cool lake bottom indicating an absence of direct volcanic interaction with lake waters
- An upper and lower thermal layer with differing CO_2 saturations
- Proximity to areas with volcanic activity

Lake Nyos, shortly after a limnic eruption

Scientists have recently determined, from investigations into the mass casualties in the 1980s at Lake Monoun and Lake Nyos, that although limnic eruptions can be indirectly related to volcanic eruptions, they are actually separate types of disaster events.

Historical Occurrences

Cattle killed by the limnic eruption at Lake Nyos

To date, this phenomenon has been observed only twice. The first was in Cameroon at Lake Monoun in 1984, causing the asphyxiation and death of 38 people living nearby. A second, deadlier eruption happened at neighbouring Lake Nyos in 1986, this time releasing over 80 million cubic metres of CO_2 and killing around 1,700 people and 3,500 livestock, again by asphyxiation.

Due to the nature of the event, it is hard to determine if limnic eruptions have happened elsewhere. A third lake, Lake Kivu, exists on the border between the Democratic Republic of the Congo and

Rwanda, and contains massive amounts of dissolved CO_2. Sample sediments from the lake were taken by professor Robert Hecky from the University of Michigan, which showed that an event caused living creatures in the lake to go extinct approximately every thousand years, and caused nearby vegetation to be swept back into the lake. Limnic eruptions can be measured on a scale using the concentration of CO_2 in the surrounding area.

The Messel pit fossil deposits of Messel, Germany, show evidence of a limnic eruption there in the early Eocene. Among the victims are perfectly preserved insects, frogs, turtles, crocodiles, birds, anteaters, insectivores, early primates and paleotheres.

Causes

For a limnic eruption to occur, the lake must be nearly saturated with gas. In the two known cas-es, the major component was CO_2; however, in Lake Kivu, scientists are concerned about the concentrations of methane gas as well. This CO_2 may come from volcanic gas emitted from under the lake or from decomposition of organic material. Before a lake is saturated, it behaves like an unopened carbonated beverage (soft drink): the CO_2 is dissolved in the water. In both the lake and the soft drink, CO_2 dissolves much more readily at higher pressure (Henry's law). This is why bubbles in a can of soda form only after the can is opened; the pressure is released and the CO_2 comes out of solution. In the case of lakes, the bottom is at a much higher pressure; the deeper it is, the higher the pressure at the bottom. This means that huge amounts of CO_2 can be dissolved in large, deep lakes. Also, CO_2 dissolves more readily in cooler water, such as that at the bottom of a lake. A small rise in water temperature can lead to the release of a large amount of CO_2.

Once the lake is saturated with CO_2, it is very unstable. A trigger is all that is needed to set off an eruption. In the case of the 1986 eruption at Lake Nyos, landslides were the suspected triggers, but an actual volcanic eruption, an earthquake, or even wind and rain storms are other possible triggers. Another possible cause of a limnic eruption is the gradual development of saturation at certain depths which can trigger the spontaneous development of gas. In any case, the trigger pushes some of the saturated water higher in the lake, where the pressure is insufficient to keep the CO_2 in solution. Bubbles start forming and the water is lifted even higher in the lake (buoyancy), where even more of the CO_2 comes out of solution. This process forms a column of gas. At this point the water at the bottom of this column is pulled up by suction, and it too loses its CO_2 in a runaway process. This eruption pours CO_2 into the air and can also displace water to form a tsunami.

There are several reasons this type of eruption is very rare. First, there must be a source of the CO_2, so regions with volcanic activity are most at risk. Second, the vast majority of lakes are holomictic, meaning their layers mix regularly. Only meromictic lakes do not mix and remain stratified, allowing CO_2 to remain dissolved. There is estimated to be only 1 meromictic lake for every 1000 holomictic lakes. Finally, a lake must be quite deep to have enough pressure to dissolve large volumes of CO_2.

Consequences

Once an eruption occurs, a large CO_2 cloud forms above the lake and expands to the neighbouring region. Because CO_2 is denser than air, it has a tendency to sink to the ground while pushing breathable air up. As a result, life forms that need to breathe oxygen suffocate once the CO_2 cloud reaches them, as there is very little oxygen in the cloud. The CO_2 can make human bodily fluids very acidic, potentially causing CO_2 poisoning. As victims gasp for air they actually hurt themselves more by inhaling the CO_2 gas.

At Lake Nyos, the gas cloud descended from the lake into a nearby village where it settled, killing nearly everyone. In this eruption, some people as far as 25 km (16 mi) from the lake died. A change in skin color on some bodies led scientists to think that the gas cloud may have contained a dissolved acid such as hydrogen chloride as well, but that hypothesis is disputed. Many victims were found with blisters on their skin. This is believed to have been caused by pressure ulcers, which are likely to have formed from the low levels of oxygen present in the blood of those asphyxiated by the carbon dioxide. Thousands of cattle and wild animals were also asphyxiated, but no official counts were made. On the other hand, vegetation nearby was mostly unaffected except for that which grew immediately adjacent to the lake. There the vegetation was damaged or destroyed by a 5-meter (16.4 ft.) tsunami from the violent eruption.

The carbon dioxide released from the lake is at a very low temperature believed to cause frostbite, which is consistent with blisters found on the skin of survivors of the incident at Lake Monoun. The survivors also reported a smell of rotten eggs and feeling warm before passing out; this is explained by the fact that at high concentrations, carbon dioxide acts as a sensory hallucinogenic. Studies with jet pilots regarding carbon dioxide report similar effects as reported by survivors, i.e. smell of rotten eggs, gunpowder and feeling of warmth.

A Possible Solution: Degassing Lakes

A siphon used by French scientists to de-gas Lake Nyos.

Efforts have been under way for several years to develop a solution to remove the gas from these lakes and prevent a build-up that could lead to another catastrophe. A team led by the French scientist Michel Halbwachs began experimenting at Lake Monoun and Lake Nyos in 1990 using siphons to degas the waters of these lakes in a controlled manner. A pipe is positioned vertically in the lake with its upper end above the water's surface. Water saturated with CO_2 enters the bottom of the pipe and rises to the top. The lower pressure at the surface allows the gas to come out of solution. Interestingly, only a small amount of water has to be mechanically pumped initially through the pipe to start the flow. As the saturated water rises, the CO_2 comes out of solution and forms bubbles. The natural buoyancy of the bubbles draws the water up the pipe at high velocity causing a large fountain at the surface. The degassifying water acts as a pump, drawing more water into the bottom of the pipe, and creating a self-sustaining flow. This is the same process that leads to a natural eruption, but in this case it is controlled by the size of the pipe.

Each pipe has a limited pumping capacity and several would be required for both Lake Monoun and Lake Nyos to degas a significant fraction of the deep lake water and render the lakes safe. The deep lake waters are slightly acidic due to the dissolved CO_2 which causes corrosion to the pipes and electronics, necessitating ongoing maintenance. There are also fears that the CO_2 from the pipes could settle on the surface of the lake forming a thin layer of unbreathable air and thus causing problems for wildlife.

In January 2001, a single pipe was installed by the French-Cameroun team on Lake Nyos and two more pipes have been installed in 2011 with funds from the United Nations Development Programme. A pipe was installed at Lake Monoun in 2003 with two more being added in 2006. These three pipes are thought to be sufficient to prevent an increase in CO_2 levels, removing approximately the same amount of gas that naturally enters at the lake bed. In January 2003, an 18-month project had been given approval to fully degas Lake Monoun, and it has since been rendered safe.

Lake Kivu's Potential Danger

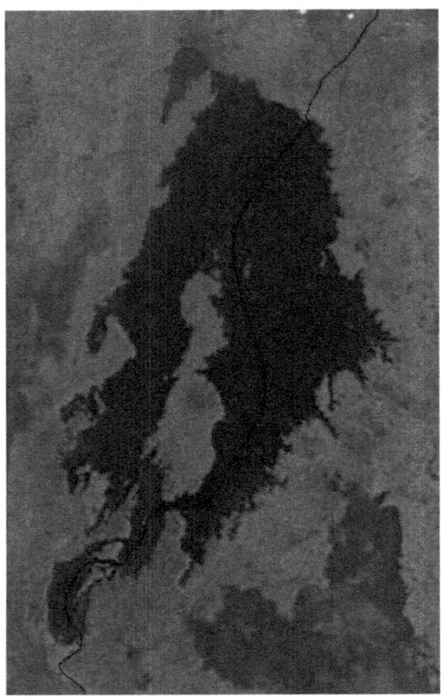

Satellite image of Lake Kivu.

Lake Kivu is not only 2,000 times larger than Lake Nyos, but is also located in a far more densely populated area, with over two million people living along its shores. It has not reached a high level of CO_2 saturation yet; if the water were to become heavily saturated, it could become an even greater risk to human and animal life. Two main changes in Lake Kivu's physical state that have brought attention to the possible limnic eruption are the high rates of methane dissociation and the rising surface temperature. Thorough research of historical and present temperature monitoring show that Lake Kivu's surface temperature is increasing by approximately 0.12 °C per decade. It is located very close to a potential trigger, Mount Nyiragongo, an active volcano that erupted in January 2002, in an active earthquake zone and close to other active volcanoes.

While the lake could be degassed in a manner similar to Lake Monoun and Lake Nyos, due to the size of the lake and the volume of gas involved such an operation would be expensive, running into millions of dollars. A scheme initiated in 2010 to utilize methane trapped in the lake as a fuel source to generate electricity in Rwanda has led to a degree of CO_2 degassing. During the procedure for extracting the flammable gas used to fuel power stations on the shore, some CO_2 is removed in a process known as catalyst scrubbing. It is unclear whether enough of the gas will be removed to eliminate the danger of a limnic eruption at Lake Kivu.

Tsunami

2011 Tōhoku earthquake and tsunami, An aerial view of damage in the Sendai region with black smoke coming from the Nippon Oil Sendai oil refinery

A *tsunami* also known as a seismic sea wave, is a series of waves in a water body caused by the displacement of a large volume of water, generally in an ocean or a large lake. Earthquakes, volcanic eruptions and other underwater explosions (including detonations of underwater nuclear devices), landslides, glacier calvings, meteorite impacts and other disturbances above or below water all have the potential to generate a tsunami. Unlike normal ocean waves which are generated by wind, or tides which are generated by the gravitational pull of the Moon and Sun, a tsunami is generated by the displacement of water.

Tsunami waves do not resemble normal undersea currents or sea waves, because their wavelength is far longer. Rather than appearing as a breaking wave, a tsunami may instead initially resemble a rapidly rising tide, and for this reason they are often referred to as *tidal waves*, although this usage is not favoured by the scientific community because tsunamis are not tidal in nature. Tsunamis generally consist of a series of waves with periods ranging from minutes to hours, arriving in a so-called "internal wave train". Wave heights of tens of metres can be generated by large events. Although the impact of tsunamis is limited to coastal areas, their destructive power can be enormous and they can affect entire ocean basins; the 2004 Indian Ocean tsunami was among the deadliest natural disasters in human history with at least 230,000 people killed or missing in 14 countries bordering the Indian Ocean.

Greek historian Thucydides suggested in his late-5th century BC *History of the Peloponnesian War*, that tsunamis were related to submarine earthquakes, but the understanding of a tsunami's nature remained slim until the 20th century and much remains unknown. Major areas of current research include trying to determine why some large earthquakes do not generate tsunamis while other smaller ones do; trying to accurately forecast the passage of tsunamis across the oceans; and also to forecast how tsunami waves interact with specific shorelines.

3D tsunami simulation

Terminology

Tsunami warning bilingual sign in Ulee Lheue, Banda Aceh in Acehnese and Indonesian

Numerous terms are used in the English language to describe waves created in a body of water by the displacement of water; however, none of the terms in frequent use are entirely accurate.

Tsunami

The term *tsunami*, meaning "harbour wave" in literal translation, comes from the Japanese, composed of the two kanji (*tsu*) meaning "harbour" and (*nami*), meaning "wave". (For the plural, one can either follow ordinary English practice and add an *s*, or use an invariable plural as in the Japanese.) While not entirely accurate, as tsunami are not restricted to harbours, *tsunami* is currently the term most widely accepted by geologists and oceanographers.

Tidal Wave

Tsunami are sometimes referred to as *tidal waves*. This once-popular term derives from the most common appearance of tsunami, which is that of an extraordinarily high tidal bore. Tsunami and tides both produce waves of water that move inland, but in the case of tsunami the inland movement of water may be much greater, giving the impression of an incredibly high and forceful tide. In recent years, the term "tidal wave" has fallen out of favour, especially in the scientific community, because tsunami actually have nothing to do with tides, which are produced by the gravitational pull of the moon and sun rather than the displacement of water. Although the meanings of "tidal" include "resembling" or "having the form or character of" the tides, use of the term *tidal wave* is discouraged by geologists and oceanographers.

Tsunami aftermath in Aceh, Indonesia, December 2004.

Seismic Sea Wave

The term *seismic sea wave* also is used to refer to the phenomenon, because the waves most often are generated by seismic activity such as earthquakes. Prior to the rise of the use of the term *tsunami* in English-speaking countries, scientists generally encouraged the use of the term *seismic sea wave* rather than the inaccurate term *tidal wave*. However, like *tsunami*, *seismic sea wave* is not a completely accurate term, as forces other than earthquakes – including underwater landslides, volcanic eruptions, underwater explosions, land or ice slumping into the ocean, meteorite impacts, and the weather when the atmospheric pressure changes very rapidly – can generate such waves by displacing water.

History

Lisbon earthquake and tsunami in 1755

While Japan may have the longest recorded history of tsunamis, the sheer destruction caused by the 2004 Indian Ocean earthquake and tsunami event mark it as the most devastating of its kind in modern times, killing around 230,000 people. The Sumatran region is not unused to tsunamis either, with earthquakes of varying magnitudes regularly occurring off the coast of the island.

Tsunamis are an often underestimated hazard in the Mediterranean Sea and parts of Europe. Of historical and current (with regard to risk assumptions) importance are the 1755 Lisbon earthquake and tsunami (which was caused by the Azores–Gibraltar Transform Fault), the 1783 Calabrian earthquakes, each causing several tens of thousands of deaths and the 1908 Messina earthquake and tsunami. The tsunami claimed more than 123,000 lives in Sicily and Calabria and is among the most deadly natural disasters in modern Europe. The Storegga Slide in the Norwegian sea and some examples of tsunamis affecting the British Isles refer to landslide and meteotsunamis predominantly and less to earthquake-induced waves.

As early as 426 BC the Greek historian Thucydides inquired in his book *History of the Peloponnesian War* about the causes of tsunami, and was the first to argue that ocean earthquakes must be the cause.

"The cause, in my opinion, of this phenomenon must be sought in the earthquake. At the point where its shock has been the most violent the sea is driven back, and suddenly recoiling with redoubled force, causes the inundation. Without an earthquake I do not see how such an accident could happen."

The Roman historian Ammianus Marcellinus (*Res Gestae* 26.10.15-19) described the typical sequence of a tsunami, including an incipient earthquake, the sudden retreat of the sea and a following gigantic wave, after the 365 AD tsunami devastated Alexandria.

Generation Mechanisms

The principal generation mechanism (or cause) of a tsunami is the displacement of a substantial volume of water or perturbation of the sea. This displacement of water is usually attributed to either earthquakes, landslides, volcanic eruptions, glacier calvings or more rarely by meteorites and nuclear tests. The waves formed in this way are then sustained by gravity. Tides do not play any part in the generation of tsunamis.

Seismicity

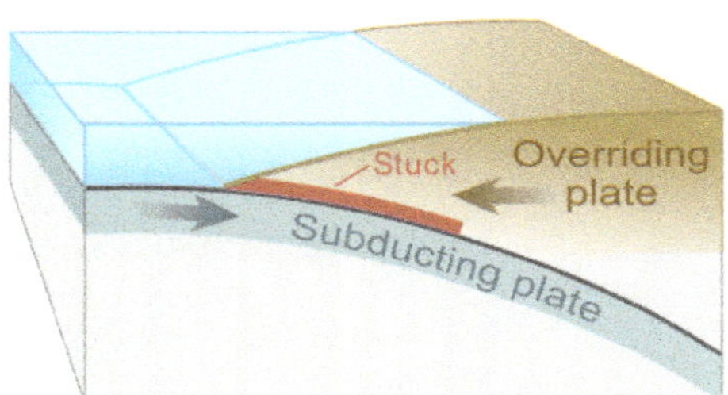

Drawing of tectonic plate boundary before earthquake

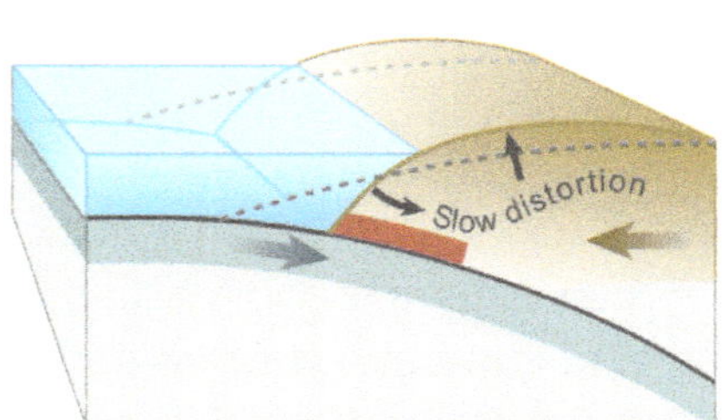

Overriding plate bulges under strain, causing tectonic uplift.

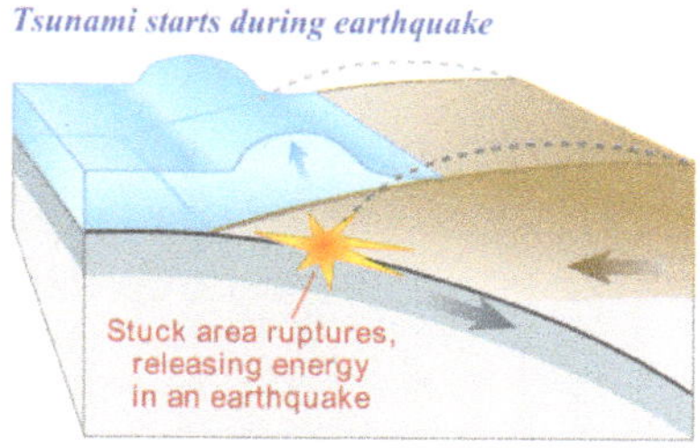

Plate slips, causing subsidence and releasing energy into water.

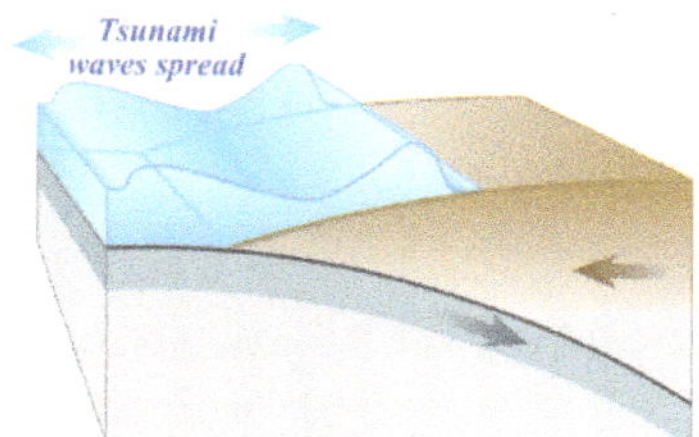

The energy released produces tsunami waves.

Tsunami can be generated when the sea floor abruptly deforms and vertically displaces the overlying water. Tectonic earthquakes are a particular kind of earthquake that are associated with the Earth's crustal deformation; when these earthquakes occur beneath the sea, the water above the deformed area is displaced from its equilibrium position. More specifically, a tsunami can be generated when thrust faults associated with convergent or destructive plate boundaries move abruptly, resulting in water displacement, owing to the vertical component of movement involved. Movement on normal (extensional) faults can also cause displacement of the seabed, but only the largest of such events (typically related to flexure in the outer trench swell) cause enough displacement to give rise to a significant tsunami, such as the 1977 Sumba and 1933 Sanriku events.

Tsunamis have a small amplitude (wave height) offshore, and a very long wavelength (often hundreds of kilometres long, whereas normal ocean waves have a wavelength of only 30 or 40 metres), which is why they generally pass unnoticed at sea, forming only a slight swell usually about 300 millimetres (12 in) above the normal sea surface. They grow in height when they reach shallower water, in a wave shoaling process described below. A tsunami can occur in any tidal state and even at low tide can still inundate coastal areas.

On April 1, 1946, a magnitude-7.8 (Richter Scale) earthquake occurred near the Aleutian Islands, Alaska. It generated a tsunami which inundated Hilo on the island of Hawai'i with a 14-metre high (46 ft) surge. The area where the earthquake occurred is where the Pacific Ocean floor is subducting (or being pushed downwards) under Alaska.

Examples of tsunami originating at locations away from convergent boundaries include Storegga about 8,000 years ago, Grand Banks 1929, Papua New Guinea 1998 (Tappin, 2001). The Grand Banks and Papua New Guinea tsunamis came from earthquakes which destabilised sediments, causing them to flow into the ocean and generate a tsunami. They dissipated before travelling transoceanic distances.

The cause of the Storegga sediment failure is unknown. Possibilities include an overloading of the sediments, an earthquake or a release of gas hydrates (methane etc.).

The 1960 Valdivia earthquake (M_w 9.5), 1964 Alaska earthquake (M_w 9.2), 2004 Indian Ocean earthquake (M_w 9.2), and 2011 Tōhoku earthquake (M_w9.0) are recent examples of powerful megathrust earthquakes that generated tsunamis (known as teletsunamis) that can cross entire oceans. Smaller (M_w 4.2) earthquakes in Japan can trigger tsunamis (called local and regional tsunamis) that can only devastate nearby coasts, but can do so in only a few minutes.

Landslides

In the 1950s, it was discovered that larger tsunamis than had previously been believed possible could be caused by giant submarine landslides. These rapidly displace large water volumes, as energy transfers to the water at a rate faster than the water can absorb. Their existence was confirmed in 1958, when a giant landslide in Lituya Bay, Alaska, caused the highest wave ever recorded, which had a height of 524 metres (over 1700 feet). The wave did not travel far, as it struck land almost immediately. Two people fishing in the bay were killed, but another boat managed to ride the wave.

Another landslide-tsunami event occurred in 1963 when a massive landslide from Monte Toc entered the Vajont Dam in Italy. The resulting wave surged over the 262 m (860 ft) high dam by 250 metres (820 ft) and destroyed several towns. Around 2,000 people died. Scientists named these waves megatsunamis.

Some geologists claim that large landslides from volcanic islands, e.g. Cumbre Vieja on La Palma in the Canary Islands, may be able to generate megatsunamis that can cross oceans, but this is disputed by many others.

In general, landslides generate displacements mainly in the shallower parts of the coastline, and there is conjecture about the nature of large landslides that enter water. This has been shown to lead to effect water in enclosed bays and lakes, but a landslide large enough to cause a transoceanic tsunami has not occurred within recorded history. Susceptible locations are believed to be the Big Island of Hawaii, Fogo in the Cape Verde Islands, La Reunion in the Indian Ocean, and Cumbre Vieja on the island of La Palma in the Canary Islands; along with other volcanic ocean islands. This is because large masses of relatively unconsolidated volcanic material occurs on the flanks and in some cases detachment planes are believed to be developing. However, there is growing controversy about how dangerous these slopes actually are.

Meteotsunamis

Some meteorological conditions, especially rapid changes in barometric pressure, as seen with the passing of a front, can displace bodies of water enough to cause trains of waves with wavelengths comparable to seismic tsunami, but usually with lower energies. These are essentially dynamically equivalent to seismic tsunami, the only differences being that meteotsunami lack the transoceanic reach of significant seismic tsunami, and that the force that displaces the water is sustained over some length of time such that meteotsunami can't be modelled as having been caused instantaneously. In spite of their lower energies, on shorelines where they can be amplified by resonance they are sometimes powerful enough to cause localised damage and potential for loss of life. They have been documented in many places, including the Great Lakes, the Aegean Sea, the English Channel, and the Balearic Islands, where they are common enough to have a local name, *rissaga*. In Sicily they are called *marubbio* and in Nagasaki Bay they are called *abiki*. Some examples of destructive meteotsunami include 31 March 1979 at Nagasaki and 15 June 2006 at Menorca, the latter causing damage in the tens of millions of euros.

Meteotsunami should not be confused with storm surges, which are local increases in sea level associated with the low barometric pressure of passing tropical cyclones, nor should they be confused with setup, the temporary local raising of sea level caused by strong on-shore winds. Storm surges and setup are also dangerous causes of coastal flooding in severe weather but their dynamics are completely unrelated to tsunami waves. They are unable to propagate beyond their sources, as waves do.

Man-made or Triggered Tsunamis

There have been studies of the potential of induction of and at least one actual attempt to create tsunami waves as a tectonic weapon.

In World War II, the New Zealand Military Forces initiated Project Seal, which attempted to create small tsunamis with explosives in the area of today's Shakespear Regional Park; the attempt failed.

There has been considerable speculation on the possibility of using nuclear weapons to cause tsunamis near to an enemy coastline. Even during World War II consideration of the idea using conventional explosives was explored. Nuclear testing in the Pacific Proving Ground by the United States seemed to generate poor results. *Operation Crossroads* fired two 20 kilotonnes of TNT (84 TJ) bombs, one in the air and one underwater, above and below the shallow (50 m (160 ft)) waters of the Bikini Atoll lagoon. Fired about 6 km (3.7 mi) from the nearest island, the waves there were no higher than 3–4 m (9.8–13.1 ft) upon reaching the shoreline. Other underwater tests, mainly *Hardtack I/Wahoo* (deep water) and *Hardtack I/Umbrella* (shallow water) confirmed the results. Analysis of the effects of shallow and deep underwater explosions indicate that the energy of the explosions doesn't easily generate the kind of deep, all-ocean waveforms which are tsunamis; most of the energy creates steam, causes vertical fountains above the water, and creates compressional waveforms. Tsunamis are hallmarked by permanent large vertical displacements of very large volumes of water which don't occur in explosions.

Characteristics

When the wave enters shallow water, it slows down and its amplitude (height) increases.

Tsunamis cause damage by two mechanisms: the smashing force of a wall of water travelling at high speed, and the destructive power of a large volume of water draining off the land and carrying a large amount of debris with it, even with waves that do not appear to be large.

The wave further slows and amplifies as it hits land. Only the largest waves crest.

While everyday wind waves have a wavelength (from crest to crest) of about 100 metres (330 ft) and a height of roughly 2 metres (6.6 ft), a tsunami in the deep ocean has a much larger wavelength of up to 200 kilometres (120 mi). Such a wave travels at well over 800 kilometres per hour (500 mph), but owing to the enormous wavelength the wave oscillation at any given point takes 20 or 30 minutes to complete a cycle and has an amplitude of only about 1 metre (3.3 ft). This makes tsunamis difficult to detect over deep water, where ships are unable to feel their passage.

The velocity of a tsunami can be calculated by obtaining the square root of the depth of the water in metres multiplied by the acceleration due to gravity (approximated to 10 m sec^2). For example, if the Pacific Ocean is considered to have a depth of 5000 metres, the velocity of a tsunami would be the square root of $\sqrt{5000 \times 10} = \sqrt{50000} = \sim 224$ metres per second (735 feet per second), which equates to a speed of ~806 kilometres per hour or about 500 miles per hour. This formula is the same as used for calculating the velocity of shallow waves, because a tsunami behaves like a shallow wave as it peak to peak value reaches from the floor of the ocean to the surface.

The reason for the Japanese name "harbour wave" is that sometimes a village's fishermen would sail out, and encounter no unusual waves while out at sea fishing, and come back to land to find their village devastated by a huge wave.

As the tsunami approaches the coast and the waters become shallow, wave shoaling compresses the wave and its speed decreases below 80 kilometres per hour (50 mph). Its wavelength diminishes to less than 20 kilometres (12 mi) and its amplitude grows enormously – in accord with Green's law. Since the wave still has the same very long period, the tsunami may take minutes to reach full height. Except for the very largest tsunamis, the approaching wave does not break, but rather appears like a fast-moving tidal bore. Open bays and coastlines adjacent to very deep water may shape the tsunami further into a step-like wave with a steep-breaking front.

When the tsunami's wave peak reaches the shore, the resulting temporary rise in sea level is termed *run up*. Run up is measured in metres above a reference sea level. A large tsunami may feature multiple waves arriving over a period of hours, with significant time between the wave crests. The first wave to reach the shore may not have the highest run up.

About 80% of tsunamis occur in the Pacific Ocean, but they are possible wherever there are large bodies of water, including lakes. They are caused by earthquakes, landslides, volcanic explosions, glacier calvings, and bolides.

Drawback

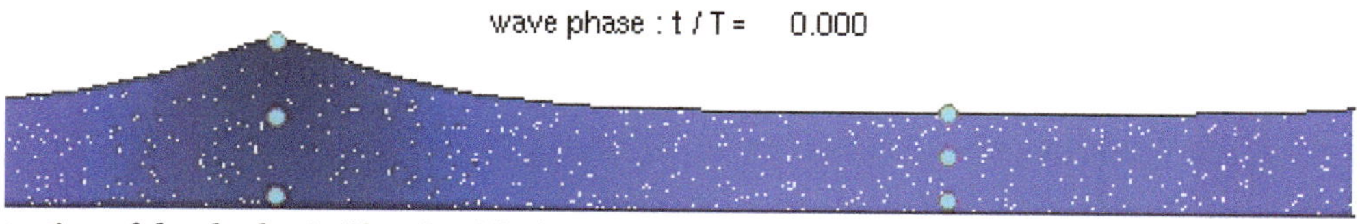

An illustration of the rhythmic "drawback" of surface water associated with a wave. It follows that a very large drawback may herald the arrival of a very large wave.

All waves have a positive and negative peak, i.e. a ridge and a trough. In the case of a propagating wave like a tsunami, either may be the first to arrive. If the first part to arrive at shore is the ridge, a massive breaking wave or sudden flooding will be the first effect noticed on land. However, if the first part to arrive is a trough, a drawback will occur as the shoreline recedes dramatically, exposing normally submerged areas. Drawback can exceed hundreds of metres, and people unaware of the danger sometimes remain near the shore to satisfy their curiosity or to collect fish from the exposed seabed.

A typical wave period for a damaging tsunami is about 12 minutes. This means that if the drawback phase is the first part of the wave to arrive, the sea will recede, with areas well below sea level exposed after 3 minutes. During the next 6 minutes the tsunami wave trough builds into a ridge, and during this time the sea is filled in and destruction occurs on land. During the next 6 minutes, the tsunami wave changes from a ridge to a trough, causing flood waters to drain and drawback to occur again. This may sweep victims and debris some distance from land. The process repeats as the next wave arrives.

Scales of Intensity and Magnitude

As with earthquakes, several attempts have been made to set up scales of tsunami intensity or magnitude to allow comparison between different events.

Intensity Scales

The first scales used routinely to measure the intensity of tsunami were the *Sieberg-Ambraseys scale*, used in the Mediterranean Sea and the *Imamura-Iida intensity scale*, used in the Pacific Ocean. The latter scale was modified by Soloviev, who calculated the Tsunami intensity *I* according to the formula

$$I = \frac{1}{2} + \log_2 H_{av} \quad \text{Pref} = \frac{1}{2}\rho v^2$$

where H_{av} is the average wave height along the nearest coast. This scale, known as the *Soloviev-Imamura tsunami intensity scale*, is used in the global tsunami catalogues compiled by the NGDC/NOAA and the Novosibirsk Tsunami Laboratory as the main parameter for the size of the tsunami.

In 2013, following the intensively studied tsunamis in 2004 and 2011, a new 12 point scale was proposed, the Integrated Tsunami Intensity Scale (ITIS-2012), intended to match as closely as possible to the modified ESI2007 and EMS earthquake intensity scales.

Magnitude Scales

The first scale that genuinely calculated a magnitude for a tsunami, rather than an intensity at a particular location was the ML scale proposed by Murty & Loomis based on the potential energy. Difficulties in calculating the potential energy of the tsunami mean that this scale is rarely used. Abe introduced the *tsunami magnitude scale M t*, calculated from,

$$M_t = a \log h + b \log R = D$$

where h is the maximum tsunami-wave amplitude (in m) measured by a tide gauge at a distance R from the epicentre, a, b and D are constants used to make the M_t scale match as closely as possible with the moment magnitude scale.

Warnings and Predictions

Tsunami warning sign

Drawbacks can serve as a brief warning. People who observe drawback (many survivors report an accompanying sucking sound), can survive only if they immediately run for high ground or seek the upper floors of nearby buildings. In 2004, ten-year-old Tilly Smith of Surrey, England, was on Maikhao beach in Phuket, Thailand with her parents and sister, and having learned about tsunamis recently in school, told her family that a tsunami might be imminent. Her parents warned others minutes before the wave arrived, saving dozens of lives. She credited her geography teacher, Andrew Kearney.

In the 2004 Indian Ocean tsunami drawback was not reported on the African coast or any other east-facing coasts that it reached. This was because the wave moved downwards on the eastern side of the fault line and upwards on the western side. The western pulse hit coastal Africa and other western areas.

A tsunami cannot be precisely predicted, even if the magnitude and location of an earthquake is known. Geologists, oceanographers, and seismologists analyse each earthquake and based on many factors may or may not issue a tsunami warning. However, there are some warning signs of an impending tsunami, and automated systems can provide warnings immediately after an earthquake in time to save lives. One of the most successful systems uses bottom pressure sensors, attached to buoys, which constantly monitor the pressure of the overlying water column.

Regions with a high tsunami risk typically use tsunami warning systems to warn the population before the wave reaches land. On the west coast of the United States, which is prone to Pacific

Ocean tsunami, warning signs indicate evacuation routes. In Japan, the community is well-educated about earthquakes and tsunamis, and along the Japanese shorelines the tsunami warning signs are reminders of the natural hazards together with a network of warning sirens, typically at the top of the cliff of surroundings hills.

The Pacific Tsunami Warning System is based in Honolulu, Hawai□i. It monitors Pacific Ocean seismic activity. A sufficiently large earthquake magnitude and other information triggers a tsunami warning. While the subduction zones around the Pacific are seismically active, not all earthquakes generate tsunami. Computers assist in analysing the tsunami risk of every earthquake that occurs in the Pacific Ocean and the adjoining land masses.

Tsunami hazard sign at Bamfield, British Columbia	A tsunami warning sign in Kamakura, Japan	The monument to the victims of the 1946 tsunami at Laupahoehoe, Hawaii

A Tsunami hazard sign (Spanish - English) in Iquique, Chile.	Tsunami memorial in Kanyakumari beach	Tsunami Evacuation Route signage along U.S. Route 101, in Washington

As a direct result of the Indian Ocean tsunami, a re-appraisal of the tsunami threat for all coastal areas is being undertaken by national governments and the United Nations Disaster Mitigation Committee. A tsunami warning system is being installed in the Indian Ocean.

One of the deep water buoys used in the DART tsunami warning system

Computer models can predict tsunami arrival, usually within minutes of the arrival time. Bottom pressure sensors can relay information in real time. Based on these pressure readings and other seismic information and the seafloor's shape (bathymetry) and coastal topography, the models estimate the amplitude and surge height of the approaching tsunami. All Pacific Rim countries collaborate in the Tsunami Warning System and most regularly practise evacuation and other procedures. In Japan, such preparation is mandatory for government, local authorities, emergency services and the population.

Some zoologists hypothesise that some animal species have an ability to sense subsonic Rayleigh waves from an earthquake or a tsunami. If correct, monitoring their behaviour could provide advance warning of earthquakes, tsunami etc. However, the evidence is controversial and is not widely accepted. There are unsubstantiated claims about the Lisbon quake that some animals escaped to higher ground, while many other animals in the same areas drowned. The phenomenon was also noted by media sources in Sri Lanka in the 2004 Indian Ocean earthquake. It is possible that certain animals (e.g., elephants) may have heard the sounds of the tsunami as it approached the coast. The elephants' reaction was to move away from the approaching noise. By contrast, some humans went to the shore to investigate and many drowned as a result.

Along the United States west coast, in addition to sirens, warnings are sent on television and radio via the National Weather Service, using the Emergency Alert System.

Forecast of Tsunami Attack Probability

Kunihiko Shimazaki (University of Tokyo), a member of Earthquake Research committee of The Headquarters for Earthquake Research Promotion of Japanese government, mentioned the plan for public announcement of tsunami attack probability forecast at Japan National Press Club on 12 May 2011. The forecast includes tsunami height, attack area and occurrence probability within 100 years ahead. The forecast would integrate the scientific knowledge of recent interdisciplinarity and aftermath of the 2011 Tōhoku earthquake and tsunami. As the plan, announcement will be available from 2014.

Mitigation

A seawall at Tsu, Japan

In some tsunami-prone countries, earthquake engineering measures have been taken to reduce the damage caused onshore.

Japan, where tsunami science and response measures first began following a disaster in 1896, has produced ever-more elaborate countermeasures and response plans. The country has built many tsunami walls of up to 12 metres (39 ft) high to protect populated coastal areas. Other localities have built floodgates of up to 15.5 metres (51 ft) high and channels to redirect the water from incoming tsunami. However, their effectiveness has been questioned, as tsunami often overtop the barriers.

The Fukushima Daiichi nuclear disaster was directly triggered by the 2011 Tōhoku earthquake and tsunami, when waves exceeded the height of the plant's sea wall. Iwate Prefecture, which is an area at high risk from tsunami, had tsunami barriers walls totalling 25 kilometres (16 mi) long at coastal towns. The 2011 tsunami toppled more than 50% of the walls and caused catastrophic damage.

The Okushiri, Hokkaidō tsunami which struck Okushiri Island of Hokkaidō within two to five minutes of the earthquake on July 12, 1993, created waves as much as 30 metres (100 ft) tall—as high as a 10-story building. The port town of Aonae was completely surrounded by a tsunami wall, but the waves washed right over the wall and destroyed all the wood-framed structures in the area. The wall may have succeeded in slowing down and moderating the height of the tsunami, but it did not prevent major destruction and loss of life.

Meteorological Disaster

Meteorological disasters are caused by extreme weather, e.g. rain, drought, snow, extreme heat or cold, ice, or wind. Violent, sudden and destructive change to the environment related to, produced by, or affecting the earth's atmosphere, especially the weather-forming processes.

Examples of weather disasters include blizzard, cyclones, droughts, hailstorms, heat waves, hurricanes, floods (caused by rain) and tornadoes.

Blizzard

Visibility reduced by blowing snow during a blizzard

A blizzard is a severe snowstorm characterized by strong sustained winds of at least 35 mph (56 km/h) and lasting for a prolonged period of time—typically three hours or more. A ground blizzard is a weather condition where snow is not falling but loose snow on the ground is lifted and blown by strong winds.

Blizzard into Tochal Skiing resort, Tehran and involved skiers.

Definition

In the United States, the National Weather Service defines a blizzard as a severe snowstorm characterized by strong winds causing blowing snow that results in low visibilities. The difference between a blizzard and a snowstorm is the strength of the wind, not the amount of snow. To be a blizzard, a snow storm must have sustained winds or frequent gusts that are greater than or equal to 56 km/h (35 mph) with blowing or drifting snow which reduces visibility to 400 m or 0.25 mi or less and must last for a prolonged period of time—typically three hours or more.

While severe cold and large amounts of drifting snow may accompany blizzards, they are not required. Blizzards can bring whiteout conditions, and can paralyze regions for days at a time, particularly where snowfall is unusual or rare.

A severe blizzard has winds over 72 km/h (45 mph), near zero visibility, and temperatures of −12 °C (10 °F) or lower. In Antarctica, blizzards are associated with winds spilling over the edge of the ice plateau at an average velocity of 160 km/h (99 mph).

Ground blizzard refers to a weather condition where loose snow or ice on the ground is lifted and blown by strong winds. The primary difference between a ground blizzard as opposed to a regular blizzard is that in a ground blizzard no precipitation is produced at the time, but rather all the precipitation is already present in the form of snow or ice at the surface.

Drifted snow near Burrow-with-Burrow, Lancashire, England, January 1963

The Australia Bureau of Meteorology describes a blizzard as, "Violent and very cold wind which is laden with snow, some part, at least, of which has been raised from snow covered ground."

Blizzard conditions of cold temperatures and strong winds can cause wind chill values that can result in hypothermia or frostbite. The wind chill factor is the amount of cooling the human body feels due to the combination of wind and temperature.

United States Storm Systems

Duluth, Minnesota blizzard, March 2007

March blizzard in North Dakota, 1966.

In the United States, storm systems powerful enough to cause blizzards usually form when the jet stream dips far to the south, allowing cold, dry polar air from the north to clash with warm, humid air moving up from the south. They are most common in the Great Plains, the Great Lakes states, and the northeastern states along the coast, and less common in the Pacific Northwest.

When cold, moist air from the Pacific Ocean moves eastward to the Rocky Mountains and the Great Plains, and warmer, moist air moves north from the Gulf of Mexico, all that is needed is a movement of cold polar air moving south to form potential blizzard conditions that may extend from the Texas Panhandle to the Great Lakes.

Conditions approaching a blizzard whiteout in Minnesota, on March 1, 2007. Note the unclear horizon near the center.

Another storm system occurs when a cold core low over the Hudson Bay area in Canada is displaced southward over southeastern Canada, the Great Lakes, and New England. When the rapidly moving cold front collides with warmer air coming north from the Gulf of Mexico, strong surface winds, significant cold air advection, and extensive wintry precipitation occur.

Low pressure systems moving out of the Rocky Mountains onto the Great Plains, a broad expanse of flat land, much of it covered in prairie, steppe and grassland, can cause thunderstorms and rain to the south and heavy snows and strong winds to the north. With few trees or other obstructions to reduce wind and blowing, this part of the country is particularly vulnerable to blizzards with very low temperatures and whiteout conditions. In a true whiteout there is no visible horizon. People can become lost in their own front yards, when the door is only 3 m (10 ft) away, and they would have to feel their way back. Motorists have to stop their cars where they are, as the road is impossible to see.

Nor'easter Blizzards

Illustration of the Great Blizzard of 1888

A nor'easter is a macro-scale storm along the East Coast of the United States and Atlantic Canada. It gets its name from the direction the wind is coming from. The usage of the term in North America comes from the wind associated with many different types of storms some of which can form in the North Atlantic Ocean and some of which form as far south as the Gulf of Mexico. The term is most often used in the coastal areas of New England and Atlantic Canada. This type of storm has characteristics similar to a hurricane. More specifically it describes a low-pressure area whose center of rotation is just off the East Coast and whose leading winds in the left-forward quadrant rotate onto land from the northeast. High storm waves may sink ships at sea and cause coastal flooding and beach erosion. Notable nor'easters include The Great Blizzard of 1888, one of the worst blizzards in U.S. history. It dropped 100–130 cm (40–50 in) of snow and had sustained winds of more than 45 miles per hour (72 km/h) that produced snowdrifts in excess of 50 feet (15 m). Railroads were shut down and people were confined to their houses for up to a week. It killed 400 people, mostly in New York.

Historic Events

1972 Iran Blizzard

The 1972 Iran Blizzard, which caused approximately 4,000 deaths, was the deadliest blizzard in recorded history. Dropping as much as 26 feet (7.9 m) of snow, it completely covered 200 villages. After a snowfall lasting nearly a week, an area the size of Wisconsin was entirely buried in snow.

The Snow Winter of 1880–1881

A snow blockade in southern Minnesota. On March 29, 1881, snowdrifts in Minnesota were higher than locomotives.

The winter of 1880–1881 is widely considered the most severe winter ever known in the United States. Many children—and their parents—learned of "The Snow Winter" through the children's book *The Long Winter* by Laura Ingalls Wilder, in which the author tells of her family's efforts to survive. The snow arrived in October 1880 and blizzard followed blizzard throughout the winter and into March 1881, leaving many areas snowbound throughout the entire winter. Accurate details in Wilder's novel include the blizzards' frequency and the deep cold, the Chicago and North Western Railway stopping trains until the spring thaw because the snow made the tracks impassable, the near-starvation of the townspeople, and the courage of her future husband Almanzo and another man, who ventured out on the open prairie in search of a cache of wheat that no one was even sure existed.

The October blizzard brought snowfalls so deep that two-story homes had snow up to the second floor windows. No one was prepared for the deep snow so early in the season and farmers all over the region were caught before their crops had even been harvested, their grain milled, or with their fuel supplies for the winter in place. By January the train service was almost entirely suspended from the region. Railroads hired scores of men to dig out the tracks but it was a wasted effort: As soon as they had finished shoveling a stretch of line, a new storm arrived, filling up the line and leaving their work useless.

There were no winter thaws and on February 2, 1881, a second massive blizzard struck that lasted for nine days. In the towns the streets were filled with solid drifts to the tops of the buildings and tunneling was needed to secure passage about town. Homes and barns were completely covered, compelling farmers to tunnel to reach and feed their stock.

When the snow finally melted in late spring of 1881, huge sections of the plains were flooded. Massive ice jams clogged the Missouri River and when they broke the downstream areas were ravaged. Most of the town of Yankton, in what is now South Dakota, was washed away when the river overflowed its banks.

The Storm of the Century

Under the weight of snow, a tree falls next to a car in Asheville, North Carolina

The Storm of the Century, also known as the Great Blizzard of 1993, was a large cyclonic storm that formed over the Gulf of Mexico on March 12, 1993, and dissipated in the North Atlantic Ocean on March 15. It is unique for its intensity, massive size and wide-reaching effect. At its height, the storm stretched from Canada towards Central America, but its main impact was on the Eastern United States and Cuba. The cyclone moved through the Gulf of Mexico, and then through the Eastern United States before moving into Canada. Areas as far south as central Alabama and Georgia received 15 to 20 cm (6 to 8 in) of snow and areas such as Birmingham, Alabama, received up to 30 cm (12 in) with isolated reports of 41 cm (16 in). Even the Florida Panhandle reported up to 10 cm (4 in), with hurricane-force wind gusts and record low barometric pressures. Between Louisiana and Cuba, hurricane-force winds produced high storm surges across northwestern Florida, which along with scattered tornadoes killed dozens of people. Record cold temperatures were seen across portions of the South and East in the wake of this storm. In the United States, the storm was responsible for the loss of electric power to over 10 million customers. It is purported to have been directly experienced by nearly 40 percent of the country's population at that time. A total of 310 people, including 10 from Cuba, perished during this storm.

List of Blizzards

North America

- The Great Snow 1717 series of four snowstorms between February 27 and March 7, 1717. There were reports of about five feet of snow already on the ground when the first of the storms hit. By the end, there were about ten feet of snow and some drifts reaching 25 feet, burying houses entirely. In the colonial era, this storm made travel impossible until the snow simply melted.

- Blizzard of 1772 (The Washington and Jefferson Snowstorm of 1772) January 26–29, 1772. One of largest D.C. and Virginia area snowstorms ever recorded. Snow accumulations of 3 feet recorded.
- "The Great Snowstorm" raged from Georgia to Maine. January 16, 1831
- Great Plains Easter Blizzard of 1873. April 13, 1873
- The Snow Winter of 1880–1881. Laura Ingles Wilder's book: The Long Winter details the effects of the blizzards in the Dakota territory of the winter of 1880–1881.
- In the three year winter period from December 1885 to March 1888, the Great Plains and Eastern United States suffered a series of the worst blizzards in this nation's history ending with the Schoolhouse Blizzard and the Great NYC Blizzard of 1888. The massive explosion of the volcano Krakatoa in the South Pacific late in August 1883 is a suspected cause of these huge blizzards during these several years. The clouds of ash it emitted continued to circulate around the world for many years. Weather patterns continued to be chaotic for years, and temperatures did not return to normal until 1888. Record rainfall was experienced in Southern California during July 1883 to June 1884. The Krakatoa eruption injected an unusually large amount of sulfur dioxide (SO2) gas high into the stratosphere which reflects sunlight and helped cool the planet over the next few years until the suspended atmospheric sulfur fell to ground.
- Plains blizzard of late 1885. In Kansas, heavy snows of late 1885 had piled drifts ten feet high.
- Kansas Blizzard of 1886. First week of January 1886. Reported that 80 percent of the cattle were frozen to death in that state alone from the cold and snow.
- Great Plains Blizzards of late 1886. On November 13, 1886 it reportedly began to began to snow and did not stop for a month in the Great Plains region.
- Great Plains Blizzard of 1887. January 9–11, 1887. Reported 72-hour blizzard that covered parts of the Great Plains in more than 16 inches of snow. Winds whipped and temperatures dropped to around -50F. So many cows that weren't killed by the cold soon died from starvation. When spring arrived, millions of the animals were dead, with around 90 percent of the open range's cattle rotting where they fell. Those present reported carcasses as far as the eye could see. Dead cattle clogged up rivers and spoiled drinking water. Many ranchers went bankrupt and others simply called it quits and moved back east. The "Great Die-Up" from the blizzard effectively concluded the romantic period of the great Plains cattle drives.
- Schoolhouse Blizzard of 1888, North American Great Plains. January 12–13, 1888. What made the storm so deadly was the timing (during work and school hours), the suddenness, and the brief spell of warmer weather that preceded it. In addition, the very strong wind fields behind the cold front and the powdery nature of the snow reduced visibilities on the open plains to zero. People ventured from the safety of their homes to do chores, go to

town, attend school, or simply enjoy the relative warmth of the day. As a result, thousands of people—including many schoolchildren—got caught in the blizzard.

- Great Blizzard of 1888 March 11–13, 1888. One of the most severe recorded blizzards in the history of the United States. On March 12, an unexpected northeaster hit New England and the mid-Atlantic, dropping up to 130 cm (50 in) of snow in the space of two days. Some 400 people died, including many sailors aboard vessels that were beset by gale-force winds and turbulent seas. In parts of New York City, snowdrifts reached up to the second story of some buildings.
- Great Blizzard of 1899 February 11–14, 1899. An extremely unusual blizzard in that it reached into the far southern states of America. It hit in February, and the area around Washington, D.C., experienced 51 hours straight of snowfall. The port of New Orleans was totally iced over; revelers participating in the New Orleans Mardi Gras had to wait for the parade routes to be shoveled free of snow. Concurrent with this blizzard was the extremely cold arctic air. Many city and state record low temperatures date back to this event, including all-time records for locations in the Midwest and South. State record lows: Nebraska reached −47 °F (−44 °C), Ohio experienced −39 °F (−39 °C), Louisiana bottomed out at −16 °F (−27 °C), and Florida dipped below zero to −2 °F (−19 °C).
- Great Lakes Storm of 1913 November 7–10, 1913. "The White Hurricane" of 1913 was the deadliest and most destructive natural disaster ever to hit the Great Lakes Basin in the Midwestern United States and the Canadian province of Ontario. It produced 140 km/h (90 mph) wind gusts, waves over 11 m (35 ft) high, and whiteout snowsqualls. It killed more than 250 people, destroyed 19 ships, and stranded 19 others. Perhaps the most well-known ship to go down in a November gale was the *SS Edmund Fitzgerald* in 1975.
- 1920 North Dakota blizzard March 15–18, 1920
- Knickerbocker Storm January 27–28, 1922
- Armistice Day Blizzard November 10–12, 1940. Took place in the Midwest region of the United States on Armistice Day. This "Panhandle hook" winter storm cut a 1,600-kilometre-wide path (1,000 mi) through the middle of the country from Kansas to Michigan. The morning of the storm was unseasonably warm but by mid afternoon conditions quickly deteriorated into a raging blizzard that would last into the next day. A total of 145 deaths were blamed on the storm, almost a third of them duck hunters who had taken time off to take advantage of the ideal hunting conditions. Weather forecasters had not predicted the severity of the oncoming storm, and as a result the hunters were not dressed for cold weather. When the storm began many hunters took shelter on small islands in the Mississippi River, and the 80 km/h (50 mph) winds and 1.5-metre (5 ft) waves overcame their encampments. Some became stranded on the islands and then froze to death in the single-digit temperatures that moved in over night. Others tried to make it to shore and drowned.
- North American blizzard of 1947 December 25–26, 1947. Was a record-breaking snowfall that began on Christmas Day and brought the Northeast United States to a standstill. It was not accompanied by high winds, but the snow fell steadily with drifts reaching 3.0 m (10 ft). Seventy-seven deaths were attributed to the blizzard.

- Great Appalachian Storm of November 1950 November 24–30, 1950
- The Mount Shasta California Snowstorm of 1959 - The storm dumped 189 inches of snow on Mount Shasta. The bulk of the snow fell on unpopulated mountainous areas, barely disrupting the residents of the Mount Shasta area. The amount of snow recorded is the largest snowfall from a single storm in North America.
- North American blizzard of 1966 January 27–31, 1966
- Chicago Blizzard of 1967 January 26–27, 1967
- February 1969 nor'easter February 8–10, 1969
- The Great Storm of 1975 known as the "Super Bowl Blizzard" or "Minnesota's Storm of the Century". January 9–12, 1975
- Groundhog Day gale of 1976 February 2, 1976
- Buffalo Blizzard of 1977 January 28-February 1, 1977. There were several feet of packed snow already on the ground, and the blizzard brought with it enough snow to reach Buffalo's record for the most snow in one season – 199.4 inches.
- Great Blizzard of 1978 also called the Cleveland Superbomb. January 25–27, 1978. Was one of the worst snowstorms the Midwest has ever seen. Wind gusts approached 160 km/h (100 mph), causing snowdrifts to reach heights of 7.6 m (25 ft) in some areas, making roadways impassable.
- Northeastern United States Blizzard of 1978 - February 6–7, 1978
- Chicago Blizzard of 1979 January 13–14, 1979
- 1991 Halloween blizzard Upper Mid-West US, October 31-November 3, 1991
- December 1992 nor'easter December 10–12, 1992
- 1993 Storm of the Century March 12–15, 1993. While the southern and eastern U.S. and Cuba received the brunt of this massive blizzard, the Storm of the Century impacted a wider area than any in recorded history.
- Blizzard of 1996 January 6–10, 1996
- April Fool's Day Blizzard March 31-April 1, 1997. US East Coast
- 1997 Western Plains winter storms October 24–26, 1997
- Mid West Blizzard of 1999 January 2–4, 1999
- January 25, 2000 Southeastern United States winter storm January 25, 2000. North Carolina and Virginia
- North American blizzard of 2003 February 14–19, 2003 (Presidents' Day Storm II)

- December 2003 nor'easter December 6–7, 2003
- North American blizzard of 2005 January 20–23, 2005
- North American blizzard of 2006 February 11–13, 2006
- Early winter 2006 North American storm complex Late November 2006
- Colorado Holiday Blizzards (2006–07) December 20–29, 2006 Colorado
- February 2007 North America blizzard February 12–20, 2007
- January 2008 North American storm complex January, 2008 West Coast US
- North American blizzard of 2008 March 6–10, 2008
- 2009 Midwest Blizzard 6–8 December 2009, a bomb cyclogenesis event that also affected parts of Canada
- North American blizzard of 2009 December 16–20, 2009
- 2009 North American Christmas blizzard December 22–28, 2009
- February 5–6, 2010 North American blizzard February 5–6, 2010 Referred to at the time as Snowmageddon was a Category 3 ("major") nor'easter and severe weather event.
- February 9–10, 2010 North American blizzard February 9–10, 2010
- February 25–27, 2010 North American blizzard February 25–27, 2010
- October 2010 North American storm complex October 23–28, 2010
- December 2010 North American blizzard December 26–29, 2010
- January 31 – February 2, 2011 North American blizzard January 31-February 2, 2011. Groundhog Day Blizzard of 2011
- 2011 Halloween nor'easter October 28-Nov 1, 2011
- Hurricane Sandy October 29–31, 2012. West Virginia, western North Carolina, and southwest Pennsylvania received heavy snowfall and blizzard conditions from this hurricane
- November 2012 nor'easter November 7–10, 2012
- December 17–22, 2012 North American blizzard December 17–22, 2012
- December 25–28, 2012 North American storm complex December 25–28, 2012
- February 2013 nor'easter February 7–20, 2013
- February 2013 Great Plains blizzard February 19-March 6, 2013
- March 2013 nor'easter March 6, 2013

- October 2013 North American storm complex October 3–5, 2013
- January 2015 North American blizzard January 26–27, 2015
- Late December 2015 North American storm complex December 26–27, 2015 Was one of the most notorious blizzards in the state of New Mexico and West Texas ever reported. It had sustained winds of over 30 mph and continuous snow precipitation that lasted over 30 hours. Dozens of vehicles were stranded in small county roads in the areas of Hobbs, Roswell, and Carlsbad New Mexico. Strong sustained winds destroyed various mobile homes.
- January 2016 United States blizzard January 20–23, 2016
- February 2016 North American storm complex February 1–8, 2016

Canada

- The Eastern Canadian Blizzard of 1971 - Dumped a foot and a half (45.7 cm) of snow on Montreal and more than two feet (61 cm) elsewhere in the region. The blizzard caused the cancellation of a Montreal Canadiens hockey game for the first time since 1918.
- Saskatchewan blizzard of 2007 - January 10, 2007 Canada

Cyclone

An extratropical cyclone near Iceland on September 4, 2003

In meteorology, a cyclone is a large scale air mass that rotates around a strong center of low atmospheric pressure. They are usually characterized by inward spiraling winds that rotate. All large-scale cyclones are centered on low-pressure areas. The largest low-pressure systems are polar vortices and extratropical cyclones of the largest scale (the synoptic scale). Warm-core cyclones such as tropical cyclones and subtropical cyclones also lie within the synoptic scale. Mesocyclones, tornadoes and dust devils lie within the smaller mesoscale. Upper level cyclones can exist without the presence of a surface low, and can pinch off from the base of the Tropical Upper Tropospheric Trough during the summer months in the Northern Hemisphere. Cyclones have also been seen on extraterrestrial planets, such as Mars and Neptune. Cyclogenesis describes the process of cyclone formation and intensification. Extratropical cyclones begin as waves in large regions of enhanced mid-latitude temperature contrasts called baroclinic zones. These zones contract and form weather fronts as the cyclonic circulation closes and intensifies. Later in their life cycle, extratropical cyclones occlude as cold air masses undercut the warmer air and become cold core systems. A cyclone's track is guided over the course of its 2 to 6 day life cycle by the steering flow of the subtropical jet stream.

Weather fronts mark the boundary between two masses of air of different temperature, humidity, and densities, and are associated with the most prominent meteorological phenomena. Strong cold fronts typically feature narrow bands of thunderstorms and severe weather, and may on occasion be preceded by squall lines or dry lines. Such fronts form west of the circulation center and generally move from west to east; warm fronts form east of the cyclone center and are usually preceded by stratiform precipitation and fog. Warm fronts move poleward ahead of the cyclone path. Occluded fronts form late in the cyclone life cycle near the center of the cyclone and often wrap around the storm center.

Tropical cyclogenesis describes the process of development of tropical cyclones. Tropical cyclones form due to latent heat driven by significant thunderstorm activity, and are warm core. Cyclones can transition between extratropical, subtropical, and tropical phases. Mesocyclones form as warm core cyclones over land, and can lead to tornado formation. Waterspouts can also form from mesocyclones, but more often develop from environments of high instability and low vertical wind shear. In the Atlantic and the northeastern Pacific oceans, a tropical cyclone is generally referred to as a hurricane (from the name of the ancient Central American deity of wind, Huracan), in the Indian and south Pacific oceans it is called a cyclone, and in the northwestern Pacific it is called a typhoon.

Structure

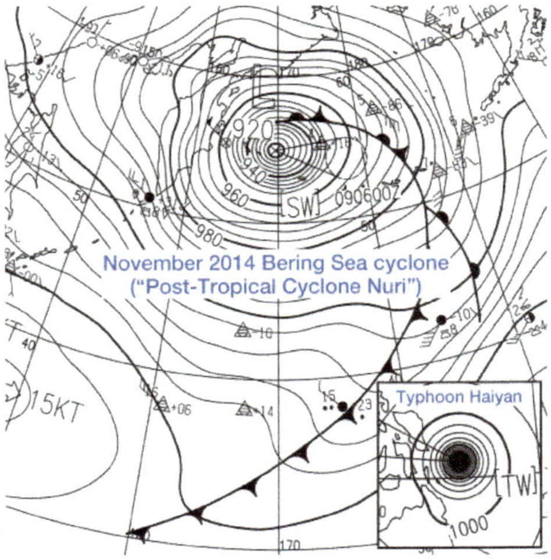

Comparison between extratropical and tropical cyclones on surface analysis

There are a number of structural characteristics common to all cyclones. A cyclone is a low-pressure area. A cyclone's center (often known in a mature tropical cyclone as the eye), is the area of lowest atmospheric pressure in the region. Near the center, the pressure gradient force (from the pressure in the center of the cyclone compared to the pressure outside the cyclone) and the force from the Coriolis effect must be in an approximate balance, or the cyclone would collapse on itself as a result of the difference in pressure.

Because of the Coriolis effect, the wind flow around a large cyclone is counterclockwise in the Northern Hemisphere and clockwise in the Southern Hemisphere. In the Northern Hemisphere, the fastest winds relative to the surface of the Earth therefore occur on the eastern side of a northward-moving cyclone and on the northern side of a westward-moving one; the opposite occurs in the Southern Hemisphere. In contrast to low pressure systems, the wind flow around high pressure systems are clockwise (anticyclonic) in the northern hemisphere, and counterclockwise in the southern hemisphere.

Formation

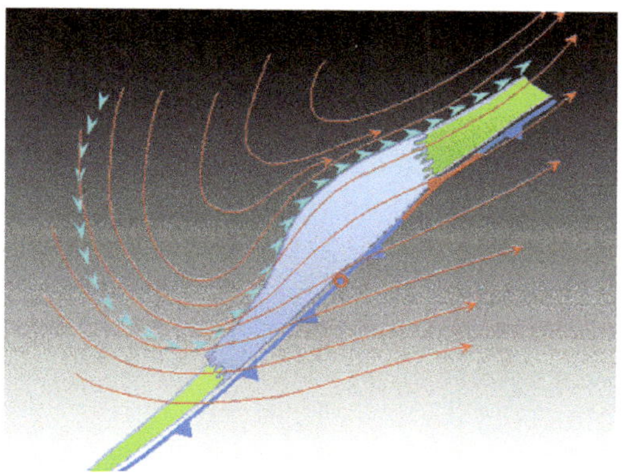

The initial extratropical low-pressure area forms at the location of the red dot on the image. It is usually perpendicular (at a right angle to) the leaf-like cloud formation seen on satellite during the early stage of cyclogenesis. The location of the axis of the upper level jet stream is in light blue.

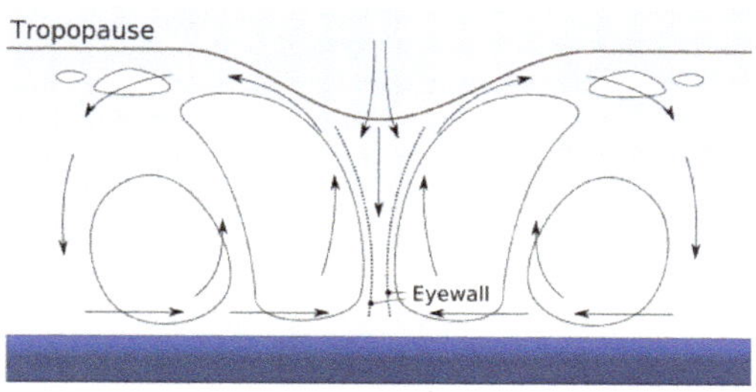

Tropical cyclones form when the energy released by the condensation of moisture in rising air causes a positive feedback loop over warm ocean waters.

Cyclogenesis is the development or strengthening of cyclonic circulation in the atmosphere. Cyclogenesis is an umbrella term for several different processes, all of which result in the development of some sort of cyclone. It can occur at various scales, from the microscale to the synoptic scale.

Extratropical cyclones begin as waves along weather fronts before occluding later in their life cycle as cold-core systems. However, some intense extratropical cyclones can become warm-core systems when a warm seclusion occurs.

Tropical cyclones form as a result of significant convective activity, and are warm core. Mesocyclones form as warm core cyclones over land, and can lead to tornado formation. Waterspouts can also form from mesocyclones, but more often develop from environments of high instability and low vertical wind shear. Cyclolysis is the opposite of cyclogenesis, and is the high-pressure system equivalent which deals with the formation of high-pressure areas—Anticyclogenesis.

A surface low can form in a variety of ways. Topography can create a surface low. Mesoscale convective systems can spawn surface lows that are initially warm core. The disturbance can grow into a wave-like formation along the front and the low will be positioned at the crest. Around the low, the flow will become cyclonic. This rotational flow will move polar air will equatorward on the west side of the low, while warm air will move poleward on the east side. A cold front will appear on the west side, while a warm front will form on the east side. Usually the cold front will move at a quicker pace than the warm front and will "catch up" with it due to the slow erosion of higher density air mass located out ahead of the cyclone. In addition, the higher density air mass sweeping in behind the cyclone strengthens the higher pressure, denser cold air mass. The cold front over takes the warm front, and reduces the length of the warm front. At this point an occluded front forms where the warm air mass is pushed upwards into a trough of warm air aloft, which is also known as a trowal.

Tropical cyclogenesis is the term that describes the development and strengthening of a tropical cyclone. The mechanisms by which tropical cyclogenesis occurs are distinctly different from those that produce mid-latitude cyclones. Tropical cyclogenesis, the development of a warm-core cyclone, begins with significant convection in a favorable atmospheric environment. There are six main requirements for tropical cyclogenesis:

1. sufficiently warm sea surface temperatures,
2. atmospheric instability,
3. high humidity in the lower to middle levels of the troposphere
4. enough Coriolis force to develop a low-pressure center
5. a preexisting low-level focus or disturbance
6. low vertical wind shear.

An average of 86 tropical cyclones of tropical storm intensity form annually worldwide, with 47 reaching hurricane/typhoon strength, and 20 becoming intense tropical cyclones (at least Category 3 intensity on the Saffir–Simpson Hurricane Scale).

Synoptic Scale

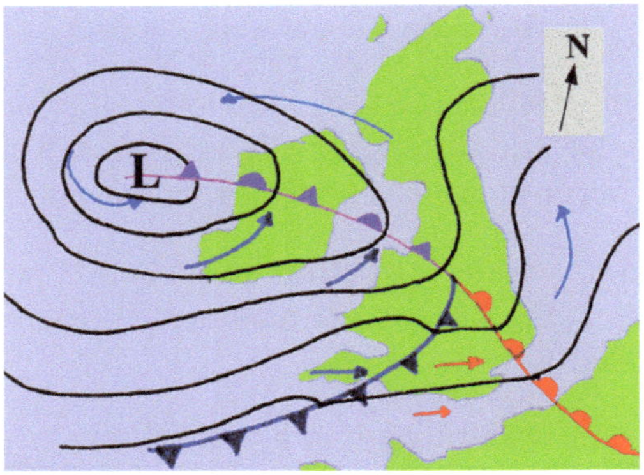

A fictitious synoptic chart of an extratropical cyclone affecting the UK and Ireland. The blue arrows between isobars indicate the direction of the wind, while the "L" symbol denotes the centre of the "low". Note the occluded, cold and warm frontal boundaries.

The following types of cyclones are identifiable in synoptic charts.

Surface-based Types

There are three main types surface-based cyclones: Extratropical cyclones, Subtropical cyclones and Tropical cyclones

Extratropical Cyclone

An extratropical cyclone is a synoptic scale low-pressure weather system that does not have tropical characteristics, as it is connected with fronts and horizontal gradients (rather than vertical) in temperature and dew point otherwise known as "baroclinic zones".

"Extratropical" is applied to cyclones outside the tropics, in the middle latitudes. These systems may also be described as "mid-latitude cyclones" due to their area of formation, or "post-tropical cyclones" when a tropical cyclone has moved (extratropical transition) beyond the tropics. They are often described as "depressions" or "lows" by weather forecasters and the general public. These are the everyday phenomena which along with anti-cyclones, drive the weather over much of the Earth.

Although extratropical cyclones are almost always classified as baroclinic since they form along zones of temperature and dewpoint gradient within the westerlies, they can sometimes become barotropic late in their life cycle when the temperature distribution around the cyclone becomes fairly uniform with radius. An extratropical cyclone can transform into a subtropical storm, and from there into a tropical cyclone, if it dwells over warm waters sufficient to warm its core, and as a result develops central convection. A particularly intense type of extratropical cyclone that strikes during winter is known colloquially as a *nor'easter*.

Polar Low

A polar low over the Sea of Japan on February 15, 2016

A polar low is a small-scale, short-lived atmospheric low-pressure system (depression) that is found over the ocean areas poleward of the main polar front in both the Northern and Southern Hemispheres. Polar lows are cold-core so they can be considered as a subset of extratropical cyclones. Polar lows were first identified on the meteorological satellite imagery that became available in the 1960s, which revealed many small-scale cloud vortices at high latitudes. The most active polar lows are found over certain ice-free maritime areas in or near the Arctic during the winter, such as the Norwegian Sea, Barents Sea, Labrador Sea and Gulf of Alaska. Polar lows dissipate rapidly when they make landfall. Antarctic systems tend to be weaker than their northern counterparts since the air-sea temperature differences around the continent are generally smaller. However, vigorous polar lows can be found over the Southern Ocean. During winter, when cold-core lows with temperatures in the mid-levels of the troposphere reach −45 °C (−49 °F) move over open waters, deep convection forms which allows polar low development to become possible. The systems usually have a horizontal length scale of less than 1,000 kilometres (620 mi) and exist for no more than a couple of days. They are part of the larger class of mesoscale weather systems. Polar lows can be difficult to detect using conventional weather reports and are a hazard to high-latitude operations, such as shipping and gas and oil platforms. Polar lows have been referred to by many other terms, such as polar mesoscale vortex, Arctic hurricane, Arctic low, and cold air depression. Today the term is usually reserved for the more vigorous systems that have near-surface winds of at least 17 m/s.

Subtropical

Subtropical Storm Andrea in 2007

A subtropical cyclone is a weather system that has some characteristics of a tropical cyclone and some characteristics of an extratropical cyclone. They can form between the equator and the 50th parallel. As early as the 1950s, meteorologists were unclear whether they should be characterized as tropical cyclones or extratropical cyclones, and used terms such as quasi-tropical and semi-tropical to describe the cyclone hybrids. By 1972, the National Hurricane Center officially recognized this cyclone category. Subtropical cyclones began to receive names off the official tropical cyclone list in the Atlantic Basin in 2002. They have broad wind patterns with maximum sustained winds located farther from the center than typical tropical cyclones, and exist in areas of weak to moderate temperature gradient.

Since they form from extratropical cyclones which have colder temperatures aloft than normally found in the tropics, the sea surface temperatures required is around 23 degrees Celsius (73 °F) for their formation, which is three degrees Celsius (5 °F) lower than for tropical cyclones. This means that subtropical cyclones are more likely to form outside the traditional bounds of the hurricane season. Although subtropical storms rarely have hurricane-force winds, they may become tropical in nature as their cores warm.

Tropical

2015 Atlantic hurricane season summary map

A tropical cyclone is a storm system characterized by a low-pressure center and numerous thunderstorms that produce strong winds and flooding rain. A tropical cyclone feeds on heat released when moist air rises, resulting in condensation of water vapour contained in the moist air. They are fueled by a different heat mechanism than other cyclonic windstorms such as nor'easters, European windstorms, and polar lows, leading to their classification as "warm core" storm systems.

Hurricane Catarina, a rare South Atlantic tropical cyclone viewed from the International Space Station on March 26, 2004

The term "tropical" refers to both the geographic origin of these systems, which form almost exclusively in tropical regions of the globe, and their dependence on Maritime Tropical air masses for their formation. The term "cyclone" refers to the storms' cyclonic nature, with counterclockwise rotation in the Northern Hemisphere and clockwise rotation in the Southern Hemisphere. Depending on their location and strength, tropical cyclones are referred to by other names, such as hurricane, typhoon, tropical storm, cyclonic storm, tropical depression, or simply as a cyclone.

While tropical cyclones can produce extremely powerful winds and torrential rain, they are also able to produce high waves and a damaging storm surge. Their winds increase the wave size, and in so doing they draw more heat and moisture into their system, thereby increasing their strength. They develop over large bodies of warm water, and hence lose their strength if they move over land. This is the reason coastal regions can receive significant damage from a tropical cyclone, while inland regions are relatively safe from strong winds. Heavy rains, however, can produce significant flooding inland. Storm surges are rises in sea level caused by the reduced pressure of the core that in effect "sucks" the water upward and from winds that in effect "pile" the water up. Storm surges can produce extensive coastal flooding up to 40 kilometres (25 mi) from the coastline. Although their effects on human populations can be devastating, tropical cyclones can also relieve drought conditions. They also carry heat and energy away from the tropics and transport it toward temperate latitudes, which makes them an important part of the global atmospheric circulation mechanism. As a result, tropical cyclones help to maintain equilibrium in the Earth's troposphere.

Many tropical cyclones develop when the atmospheric conditions around a weak disturbance in the atmosphere are favorable. Others form when other types of cyclones acquire tropical characteristics. Tropical systems are then moved by steering winds in the troposphere; if the conditions remain favorable, the tropical disturbance intensifies, and can even develop an eye. On the other end of the spectrum, if the conditions around the system deteriorate or the tropical cyclone makes landfall, the system will weaken and eventually dissipate. A tropical cyclone can become extratropical as it moves toward higher latitudes if its energy source changes from heat released by condensation to differences in temperature between air masses. A tropical cyclone is usually not considered to become subtropical during its extratropical transition.

Atlantic hurricane	Pacific typhoon	North Indian Ocean cyclone	South Pacific cyclone

Upper Level Types

Polar Cyclone

A polar, sub-polar, or Arctic cyclone (also known as a polar vortex) is a vast area of low pressure which strengthens in the winter and weakens in the summer. A polar cyclone is a low-pressure weather system, usually spanning 1,000 kilometres (620 mi) to 2,000 kilometres (1,200 mi), in which the air circulates in a counterclockwise direction in the northern hemisphere, and a clockwise direction in the southern hemisphere. The Coriolis acceleration acting on the air masses moving poleward at high altitude, causes a counterclockwise circulation at high altitude. The poleward movement of air originates from the air circulation of the Polar cell. The polar low is not driven by convection as are tropical cyclones, nor the cold and warm air mass interactions as are extratropical cyclones, but is an artifact of the global air movement of the Polar cell. The base of the polar low is in the mid to upper troposphere. In the Northern Hemisphere, the polar cyclone has two centers on average. One center lies near Baffin Island and the other over northeast Siberia. In the southern hemisphere, it tends to be located near the edge of the Ross ice shelf near 160 west longitude. When the polar vortex is strong, its effect can be felt at the surface as a westerly wind (toward the east). When the polar cyclone is weak, significant cold outbreaks occur.

TUTT Cell

Under specific circumstances, upper level cold lows can break off from the base of the Tropical Upper Tropospheric Trough (TUTT), which is located mid-ocean in the Northern Hemisphere during the summer months. These upper tropospheric cyclonic vortices, also known as TUTT cells or TUTT lows, usually move slowly from east-northeast to west-southwest, and their bases generally do not extend below 20,000 feet in altitude. A weak inverted surface trough within the trade wind is generally found underneath them, and they may also be associated with broad areas of high-level clouds. Downward development results in an increase of cumulus clouds and the appearance of a surface vortex. In rare cases, they become warm-core tropical cyclones. Upper cyclones and the upper troughs which trail tropical cyclones can cause additional outflow channels and aid in their intensification. Developing tropical disturbances can help create or deepen upper troughs or upper lows in their wake due to the outflow jet emanating from the developing tropical disturbance/cyclone.

Mesoscale

The following types of cyclones are not identifiable in synoptic charts.

Mesocyclone

A mesocyclone is a vortex of air, 2.0 kilometres (1.2 mi) to 10 kilometres (6.2 mi) in diameter (the mesoscale of meteorology), within a convective storm. Air rises and rotates around a vertical axis, usually in the same direction as low-pressure systems in both northern and southern hemisphere. They are most often cyclonic, that is, associated with a localized low-pressure region within a supercell. Such storms can feature strong surface winds and severe hail. Mesocyclones often occur together with updrafts in supercells, where tornadoes may form. About 1700 mesocyclones form annually across the United States, but only half produce tornadoes.

Tornado

A tornado is a violently rotating column of air that is in contact with both the surface of the earth and a cumulonimbus cloud or, in rare cases, the base of a cumulus cloud. Also referred to as twisters, a collequial term in America, or cyclones, although the word cyclone is used in meteorology, in a wider sense, to name any closed low-pressure circulation.

Dust Devil

A dust devil is a strong, well-formed, and relatively long-lived whirlwind, ranging from small (half a metre wide and a few metres tall) to large (more than 10 metres wide and more than 1000 metres tall). The primary vertical motion is upward. Dust devils are usually harmless, but can on rare occasions grow large enough to pose a threat to both people and property.

Waterspout

A waterspout is a columnar vortex forming over water that is, in its most common form, a non-supercell tornado over water that is connected to a cumuliform cloud. While it is often weaker than most of its land counterparts, stronger versions spawned by mesocyclones do occur.

Steam Devil

A gentle vortex over calm water or wet land made visible by rising water vapour.

Fire Whirl

A fire whirl – also colloquially known as a fire devil, fire tornado, firenado, or fire twister – is a whirlwind induced by a fire and often made up of flame or ash.

Climate Change

Scientists warn that climate change could increase the intensity of typhoons as climate change projections show that the difference in temperature between the ocean – the heat source for cyclones – and the storm tops – the cold parts of cyclones – are likely to increase. Climate change is predicted to increase the frequency of high-intensity storms in selected ocean basins. While the effect changing climate is having on tropical storms remains largely unresolved scientists and president of Vanuatu Baldwin Lonsdale say the devastation caused by Pam, was aggravated by climate change.

Other Planets

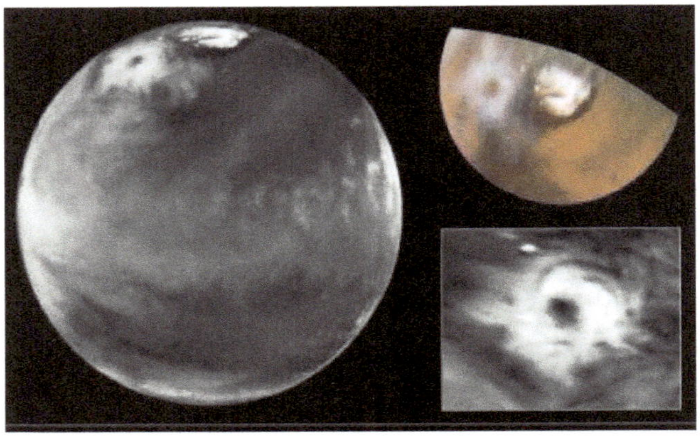

Cyclone on Mars, imaged by the Hubble Space Telescope

Cyclones are not unique to Earth. Cyclonic storms are common on Jovian planets, such as the Small Dark Spot on Neptune. It is about one third the diameter of the Great Dark Spot and received the nickname "Wizard's Eye" because it looks like an eye. This appearance is caused by a white cloud in the middle of the Wizard's Eye. Mars has also exhibited cyclonic storms. Jovian storms like the Great Red Spot are usually mistakenly named as giant hurricanes or cyclonic storms. However, this is inaccurate, as the Great Red Spot is, in fact, the inverse phenomenon, an anticyclone.

Drought

Contraction/Desiccation cracks in dry earth (Sonoran desert, Mexico).

A drought is a period of below-average precipitation in a given region, resulting in prolonged shortages in its water supply, whether atmospheric, surface water or ground water. A drought can last for months or years, or may be declared after as few as 15 days. It can have a substantial impact on the ecosystem and agriculture of the affected region and harm to the local economy. Annual dry seasons in the tropics significantly increase the chances of a drought developing and subsequent bush fires. Periods of heat can significantly worsen drought conditions by hastening evaporation of water vapour.

Many plant species, such as those in the family Cactaceae (or cacti), have drought tolerance adaptations like reduced leaf area and waxy cuticles to enhance their ability to tolerate drought. Some others survive dry periods as buried seeds. Semi-permanent drought produces arid biomes such as deserts and grasslands. Prolonged droughts have caused mass migrations and humanitarian crises. Most arid ecosystems have inherently low productivity. The most prolonged drought ever in the world in recorded history occurred in the Atacama Desert in Chile (400 Years).

Causes of Drought

Precipitation Deficiency

Ancient Meso-American civilizations may have amplified droughts by deforestation.

Mechanisms of producing precipitation include convective, stratiform, and orographic rainfall. Convective processes involve strong vertical motions that can cause the overturning of the atmosphere in that location within an hour and cause heavy precipitation, while stratiform processes involve weaker upward motions and less intense precipitation over a longer duration. Precipitation can be divided into three categories, based on whether it falls as liquid water, liquid water that freezes on contact with the surface, or ice. Drought are mainly course by in low rain areas. If these factors do not support precipitation volumes sufficient to reach the surface over a sufficient time, the result is a drought. Drought can be triggered by a high level of reflected sunlight and above average prevalence of high pressure systems, winds carrying continental, rather than oceanic air masses, and ridges of high pressure areas aloft can prevent or restrict the developing of thunderstorm activity or rainfall over one certain region. Once a region is within drought, feedback mechanisms such as local arid air, hot conditions which can promote warm core ridging, and minimal evapotranspiration can worsen drought conditions.

Dry Season

Within the tropics, distinct, wet and dry seasons emerge due to the movement of the Intertropical Convergence Zone or Monsoon trough. The dry season greatly increases drought occurrence, and is characterized by its low humidity, with watering holes and rivers drying up. Because of the lack of these watering holes, many grazing animals are forced to migrate due to the lack of water and feed to more fertile spots. Examples of such animals are zebras, elephants, and wildebeest. Because of the lack of water in the plants, bushfires are common. Since water vapor becomes more energetic with increasing temperature, more water vapor is required to increase relative humidity values to 100% at higher temperatures (or to get the temperature to fall to the dew point). Periods of warmth quicken the pace of fruit and vegetable production, increase evaporation and transpiration from plants, and worsen drought conditions.

Sheep on a drought affected paddock near Uranquinty, New South Wales.

El Niño

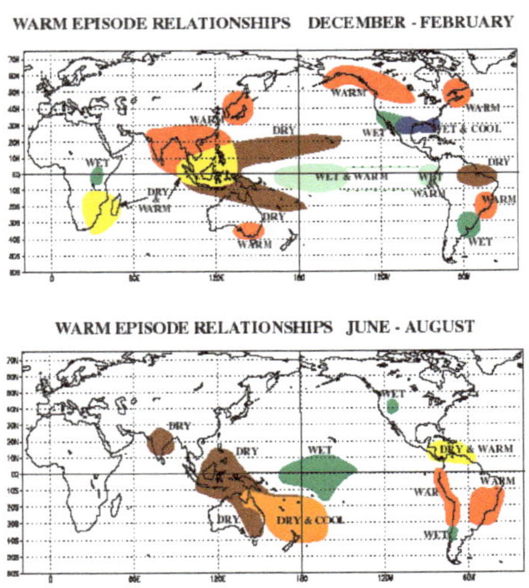

Regional impacts of warm ENSO episodes (El Niño)

Drier and hotter weather occurs in parts of the Amazon River Basin, Colombia, and Central America during El Niño events. Winters during the El Niño are warmer and drier than average conditions in the Northwest, northern Midwest, and northern Mideast United States, so those regions experience reduced snowfalls. Conditions are also drier than normal from December to February in south-central Africa, mainly in Zambia, Zimbabwe, Mozambique, and Botswana. Direct effects of El Niño resulting in drier conditions occur in parts of Southeast Asia and Northern Australia, increasing bush fires, worsening haze, and decreasing air quality dramatically. Drier-than-normal conditions are also in general observed in Queensland, inland Victoria, inland New South Wales, and eastern Tasmania from June to August. As warm water spreads from the west Pacific and the Indian Ocean to the east Pacific, it causes extensive drought in the western Pacific. Singapore experienced the driest February in 2014 since records began in 1869, with only 6.3 mm of rain falling in the month and temperatures hitting as high as 35 °C on 26 February. The years 1968 and 2005 had the next driest Februaries, when 8.4 mm of rain fell.

Erosion and Human Activities

Fires on Borneo and Sumatra, 2006. People use slash-and-burn deforestation to clear land for agriculture.

Human activity can directly trigger exacerbating factors such as over farming, excessive irrigation, deforestation, and erosion adversely impact the ability of the land to capture and hold water. In arid climates, the main source of erosion is wind. Erosion can be the result of material movement by the wind. The wind can cause small particles to be lifted and therefore moved to another region (deflation). Suspended particles within the wind may impact on solid objects causing erosion by abrasion (ecological succession). Wind erosion generally occurs in areas with little or no vegetation, often in areas where there is insufficient rainfall to support vegetation.

Fields outside Benambra, Victoria, Australia suffering from drought conditions.

Loess is a homogeneous, typically nonstratified, porous, friable, slightly coherent, often calcareous, fine-grained, silty, pale yellow or buff, windblown (Aeolian) sediment. It generally occurs as a widespread blanket deposit that covers areas of hundreds of square kilometers and tens of meters thick. Loess often stands in either steep or vertical faces. Loess tends to develop into highly rich soils. Under appropriate climatic conditions, areas with loess are among the most agriculturally productive in the world. Loess deposits are geologically unstable by nature, and will erode very readily. Therefore, windbreaks (such as big trees and bushes) are often planted by farmers to reduce the wind erosion of loess. Wind erosion is much more severe in arid areas and during times of drought. For example, in the Great Plains, it is estimated that soil loss due to wind erosion can be as much as 6100 times greater in drought years than in wet years.

Climate Change

Activities resulting in global climate change are expected to trigger droughts with a substantial impact on agriculture throughout the world, and especially in developing nations. Overall, global warming will result in increased world rainfall. Along with drought in some areas, flooding and erosion will increase in others. Paradoxically, some proposed solutions to global warming that focus on more active techniques, solar radiation management through the use of a space sunshade for one, may also carry with them increased chances of drought.

Types

As a drought persists, the conditions surrounding it gradually worsen and its impact on the local population gradually increases. People tend to define droughts in three main ways:

1. Meteorological drought is brought about when there is a prolonged time with less than average precipitation. Meteorological drought usually precedes the other kinds of drought.
2. Agricultural droughts are droughts that affect crop production or the ecology of the range. This condition can also arise independently from any change in precipitation levels when soil conditions and erosion triggered by poorly planned agricultural endeavors cause a shortfall in water available to the crops. However, in a traditional drought, it is caused by an extended period of below average precipitation.
3. Hydrological drought is brought about when the water reserves available in sources such as aquifers, lakes and reservoirs fall below the statistical average. Hydrological drought tends to show up more slowly because it involves stored water that is used but not replenished. Like an agricultural drought, this can be triggered by more than just a loss of rainfall. For instance, Kazakhstan was recently[when?] awarded a large amount of money by the World Bank to restore water that had been diverted to other nations from the Aral Sea under Soviet rule. Similar circumstances also place their largest lake, Balkhash, at risk of completely drying out.

Consequences of Drought

A Mongolian gazelle dead due to drought.

The effects of droughts and water shortages can be divided into three groups: environmental, economic and social consequences. In the case of environmental effects: lower surface and subterranean water levels, lower flow levels (with a decrease below the minimum leading to direct danger for amphibian life), increased pollution of surface water, the drying out of wetlands, more and larger fires, higher deflation intensity, losing biodiversity, worse health of trees and the appearance of pests and den droid diseases. Economic losses include lower agricultural, forest, game and fishing output, higher food production costs, lower energy production levels in hydro plants, losses caused by depleted water tourism and transport revenue, problems with water supply for the energy sector and technological processes in metallurgy, mining, the chemical, paper, wood, foodstuff industries etc., disruption of water supplies for municipal economies. Meanwhile, social costs include the negative effect on the health of people directly exposed to this phenomenon (excessive heat waves), possible limitation of water supplies and its increased pollution levels, high food costs, stress caused by failed harvests, etc. This is why droughts and fresh water shortages may be considered as a factor which increases the gap between developed and developing countries.

The effect varies according to vulnerability. For example, subsistence farmers are more likely to migrate during drought because they do not have alternative food sources. Areas with populations that depend on water sources as a major food source are more vulnerable to famine.

Drought can also reduce water quality, because lower water flows reduce dilution of pollutants and increase contamination of remaining water sources. Common consequences of drought include:

- Diminished crop growth or yield productions and carrying capacity for livestock
- Dust bowls, themselves a sign of erosion, which further erode the landscape
- Dust storms, when drought hits an area suffering from desertification and erosion
- Famine due to lack of water for irrigation
- Habitat damage, affecting both terrestrial and aquatic wildlife
- Hunger, drought provides too little water to support food crops.
- Malnutrition, dehydration and related diseases
- Mass migration, resulting in internal displacement and international refugees
- Reduced electricity production due to reduced water flow through hydroelectric dams
- Shortages of water for industrial users
- Snake migration, which results in snakebites
- Social unrest
- War over natural resources, including water and food
- Wildfires, such as Australian bushfires, are more common during times of drought and even death of people.
- Exposure and oxidation of acid sulfate soils due to falling surface and groundwater levels.

- Cyanotoxin accumulation within food chains and water supply, some of which are among the most potent toxins known to science, can cause cancer with low exposure over long term. High levels of microcystin has been found in San Francisco Bay Area salt water shellfish and fresh water supplies throughout the state of California in 2016.

Globally

Drought is a normal, recurring feature of the climate in most parts of the world. It is among the earliest documented climatic events, present in the Epic of Gilgamesh and tied to the biblical story of Joseph's arrival in and the later Exodus from Ancient Egypt. Hunter-gatherer migrations in 9,500 BC Chile have been linked to the phenomenon, as has the exodus of early humans out of Africa and into the rest of the world around 135,000 years ago.

A South Dakota farm during the Dust Bowl, 1936

Examples

Well-known historical droughts include:

- 1900 India killing between 250,000 and 3.25 million.
- 1921–22 Soviet Union in which over 5 million perished from starvation due to drought
- 1928–30 Northwest China resulting in over 3 million deaths by famine.
- 1936 and 1941 Sichuan Province China resulting in 5 million and 2.5 million deaths respectively.
- The 1997–2009 Millennium Drought in Australia led to a water supply crisis across much of the country. As a result, many desalination plants were built for the first time.
- In 2006, Sichuan Province China experienced its worst drought in modern times with nearly 8 million people and over 7 million cattle facing water shortages.
- 12-year drought that was devastating southwest Western Australia, southeast South Australia, Victoria and northern Tasmania was "very severe and without historical precedent".

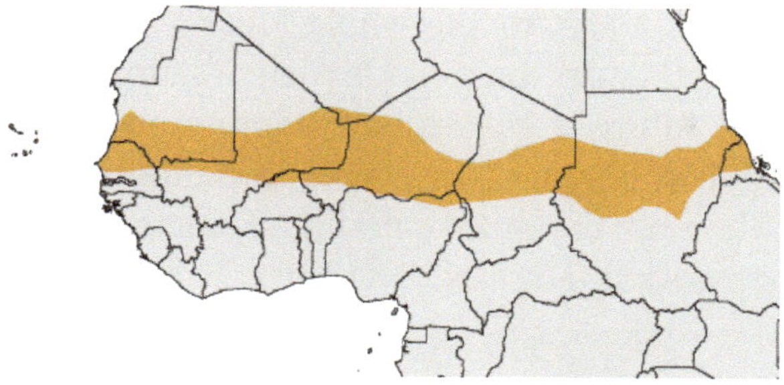

Affected areas in the western Sahel belt during the 2012 drought.

The Darfur conflict in Sudan, also affecting Chad, was fueled by decades of drought; combination of drought, desertification and overpopulation are among the causes of the Darfur conflict, because the Arab Baggara nomads searching for water have to take their livestock further south, to land mainly occupied by non-Arab farming people.

Approximately 2.4 billion people live in the drainage basin of the Himalayan rivers. India, China, Pakistan, Bangladesh, Nepal and Myanmar could experience floods followed by droughts in coming decades. Drought in India affecting the Ganges is of particular concern, as it provides drinking water and agricultural irrigation for more than 500 million people. The west coast of North America, which gets much of its water from glaciers in mountain ranges such as the Rocky Mountains and Sierra Nevada, also would be affected.

Drought affected area in Karnataka, India in 2012.

In 2005, parts of the Amazon basin experienced the worst drought in 100 years. A 23 July 2006 article reported Woods Hole Research Center results showing that the forest in its present form could survive only three years of drought. Scientists at the Brazilian National Institute of Amazonian Research argue in the article that this drought response, coupled with the effects of deforestation on regional climate, are pushing the rainforest towards a "tipping point" where it would irreversibly start to die. It concludes that the rainforest is on the brink of being turned into savanna or desert, with catastrophic consequences for the world's climate. According to the WWF, the combination of climate change and deforestation increases the drying effect of dead trees that fuels forest fires.

By far the largest part of Australia is desert or semi-arid lands commonly known as the outback. A 2005 study by Australian and American researchers investigated the desertification of the interior, and suggested that one explanation was related to human settlers who arrived about 50,000 years ago. Regular burning by these settlers could have prevented monsoons from reaching interior Australia. In June 2008 it became known that an expert panel had warned of long term, maybe irreversible, severe ecological damage for the whole Murray-Darling basin if it did not receive sufficient water by October 2008. Australia could experience more severe droughts and they could become more frequent in the future, a government-commissioned report said on July 6, 2008. Australian environmentalist Tim Flannery, predicted that unless it made drastic changes, Perth in Western Australia could become the world's first ghost metropolis, an abandoned city with no more water to sustain its population. The long Australian Millennial drought broke in 2010.

Lake Chad in a 2001 satellite image. The lake has shrunk by 95% since the 1960s.

Recurring droughts leading to desertification in East Africa have created grave ecological catastrophes, prompting food shortages in 1984–85, 2006 and 2011. During the 2011 drought, an estimated 50,000 to 150,000 people were reported to have died, though these figures and the extent of the crisis are disputed. In February 2012, the UN announced that the crisis was over due to a scaling up of relief efforts and a bumper harvest. Aid agencies subsequently shifted their emphasis to recovery efforts, including digging irrigation canals and distributing plant seeds.

In 2012, a severe drought struck the western Sahel. The Methodist Relief & Development Fund (MRDF) reported that more than 10 million people in the region were at risk of famine due to a month-long heat wave that was hovering over Niger, Mali, Mauritania and Burkina Faso. A fund of about £20,000 was distributed to the drought-hit countries.

Protection, Mitigation and Relief

Succulent plants are well-adapted to survive long periods of drought.

Water distribution on Marshall Islands during El Niño.

Agriculturally, people can effectively mitigate much of the impact of drought through irrigation and crop rotation. Failure to develop adequate drought mitigation strategies carries a grave human cost in the modern era, exacerbated by ever-increasing population densities. President Roosevelt on April 27, 1935, signed documents creating the Soil Conservation Service (SCS)—now the Natural Resources Conservation Service (NRCS). Models of the law were sent to each state where they were enacted. These were the first enduring practical programs to curtail future susceptibility to drought, creating agencies that first began to stress soil conservation measures to protect farm lands today. It was not until the 1950s that there was an importance placed on water conservation was put into the existing laws (NRCS 2014).

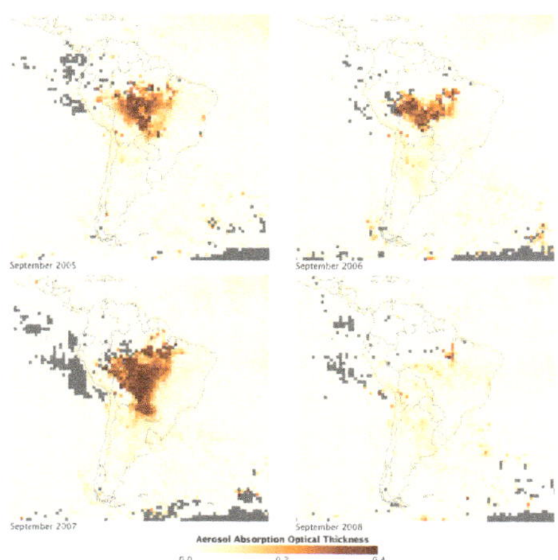

Aerosols over the Amazon each September for four burning seasons (2005 through 2008) during the Amazon basin drought. The aerosol scale (yellow to dark reddish-brown) indicates the relative amount of particles that absorb sunlight.

Strategies for drought protection, mitigation or relief include:

- Dams - many dams and their associated reservoirs supply additional water in times of drought.

- Cloud seeding - a form of intentional weather modification to induce rainfall. This remains a hotly debated topic, as the United States National Research Council released a report in 2004 stating that to date, there is still no convincing scientific proof of the efficacy of intentional weather modification.
- Desalination - of sea water for irrigation or consumption.
- Drought monitoring - Continuous observation of rainfall levels and comparisons with current usage levels can help prevent man-made drought. For instance, analysis of water usage in Yemen has revealed that their water table (underground water level) is put at grave risk by over-use to fertilize their Khat crop. Careful monitoring of moisture levels can also help predict increased risk for wildfires, using such metrics as the Keetch-Byram Drought Index or Palmer Drought Index.
- Land use - Carefully planned crop rotation can help to minimize erosion and allow farmers to plant less water-dependent crops in drier years.
- Outdoor water-use restriction - Regulating the use of sprinklers, hoses or buckets on outdoor plants, filling pools, and other water-intensive home maintenance tasks. Xeriscaping yards can significantly reduce unnecessary water use by residents of towns and cities.
- Rainwater harvesting - Collection and storage of rainwater from roofs or other suitable catchments.
- Recycled water - Former wastewater (sewage) that has been treated and purified for reuse.
- Transvasement - Building canals or redirecting rivers as massive attempts at irrigation in drought-prone areas.

Thunderstorm

A typical thunderstorm over a field

A thunderstorm, also known as an electrical storm, lightning storm, or thundershower, is a storm characterized by the presence of lightning and its acoustic effect on the Earth's atmosphere, known as thunder. Thunderstorms occur in association with a type of cloud known as a cumulonimbus. They are usually accompanied by strong winds, heavy rain, and sometimes snow, sleet, hail, or, in contrast, no precipitation at all. Thunderstorms may line up in a series or become a rainband, known as a squall line. Strong or severe thunderstorms, known as supercells, rotate as do cyclones.

While most thunderstorms move with the mean wind flow through the layer of the troposphere that they occupy, vertical wind shear sometimes causes a deviation in their course at a right angle to the wind shear direction.

Warm, moist updraft from a thunderstorm associated with a southward-moving frontal boundary - taken from Texarkana, Texas looking north.

Thunderstorms result from the rapid upward movement of warm, moist air, sometimes along a front. As the warm, moist air moves upward, it cools, condenses, and forms a cumulonimbus cloud that can reach heights of over 20 kilometres (12 mi). As the rising air reaches its dew point temperature, water vapor condenses into water droplets or ice, reducing pressure locally within the thunderstorm cell. Any precipitation falls the long distance through the clouds towards the Earth's surface. As the droplets fall, they collide with other droplets and become larger. The falling droplets create a downdraft as it pulls cold air with it, and this cold air spreads out at the Earth's surface, occasionally causing strong winds that are commonly associated with thunderstorms.

Thunderstorms can form and develop in any geographic location but most frequently within the mid-latitude, where warm, moist air from tropical latitudes collides with cooler air from polar latitudes. Thunderstorms are responsible for the development and formation of many severe weather phenomena. Thunderstorms, and the phenomena that occur along with them, pose great hazards. Damage that results from thunderstorms is mainly inflicted by downburst winds, large hailstones, and flash flooding caused by heavy precipitation. Stronger thunderstorm cells are capable of producing tornadoes and waterspouts.

There are four types of thunderstorms: single-cell, multi-cell cluster, multi-cell lines, and supercells. Supercell thunderstorms are the strongest and the most associated with severe weather phenomena. Mesoscale convective systems formed by favorable vertical wind shear within the tropics and subtropics can be responsible for the development of hurricanes. Dry thunderstorms, with no precipitation, can cause the outbreak of wildfires from the heat generated from the cloud-to-ground lightning that accompanies them. Several means are used to study thunderstorms: weather radar, weather stations, and video photography. Past civilizations held various myths concerning thunderstorms and their development as late as the 18th century. Beyond the Earth's atmosphere, thunderstorms have also been observed on the planets of Jupiter, Saturn, Neptune, and Venus.

Life Cycle

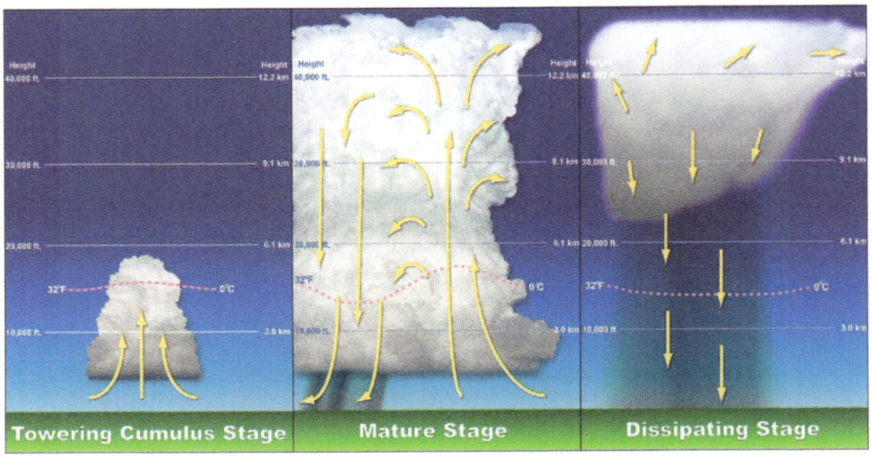

Stages of a thunderstorm's life.

Warm air has a lower density than cool air, so warmer air rises upwards and cooler air will settle at the bottom (this effect can be seen with a hot air balloon). Clouds form as relatively warmer air, carrying moisture, rises within cooler air. The moist air rises, and, as it does so, it cools and some of the water vapor in that rising air condenses. When the moisture condenses, it releases energy known as latent heat of vaporization, which allows the rising packet of air to cool less than the cooler surrounding air continuing the cloud's ascension. If enough instability is present in the atmosphere, this process will continue long enough for cumulonimbus clouds to form and produce lightning and thunder. Meteorological indices such as convective available potential energy (CAPE) and the lifted index can be used to assist in determining potential upward vertical development of clouds. Generally, thunderstorms require three conditions to form:

1. Moisture
2. An unstable airmass
3. A lifting force (heat)

All thunderstorms, regardless of type, go through three stages: the developing stage, the mature stage, and the dissipation stage. The average thunderstorm has a 24 km (15 mi) diameter. Depending on the conditions present in the atmosphere, each of these three stages take an average of 30 minutes.

Cumulus Stage

The first stage of a thunderstorm is the cumulus stage or developing stage. During this stage, masses of moisture are lifted upwards into the atmosphere. The trigger for this lift can be solar illumination, where the heating of the ground produces thermals, or where two winds converge forcing air upwards, or where winds blow over terrain of increasing elevation. The moisture carried upward cools into liquid drops of water due to lower temperatures at high altitude, which appear as *cumulus* clouds. As the water vapor condenses into liquid, latent heat is released, which warms the air, causing it to become less dense than the surrounding, drier air. The air tends to rise in an *updraft* through the process of convection (hence the term convective precipitation). This process creates a low-pressure zone within and beneath the forming thunderstorm. In a typical thunderstorm, approximately 5×10^8 kg of water vapor are lifted into the Earth's atmosphere.

Mature Stage

Anvil-shaped thundercloud in the mature stage over Swifts Creek, Victoria

In the mature stage of a thunderstorm, the warmed air continues to rise until it reaches an area of warmer air and can rise no farther. Often this 'cap' is the tropopause. The air is instead forced to spread out, giving the storm a characteristic anvil shape. The resulting cloud is called *cumulonimbus incus*. The water droplets coalesce into larger and heavier droplets and freeze to become ice particles. As these fall, they melt to become rain. If the updraft is strong enough, the droplets are held aloft long enough to become so large that they do not melt completely but fall as hail. While updrafts are still present, the falling rain drags the surrounding air with it, creating *downdrafts* as well. The simultaneous presence of both an updraft and a downdraft marks the mature stage of the storm and produces cumulonimbus clouds. During this stage, considerable internal turbulence can occur within, which manifests as strong winds, severe lightning, and even tornadoes.

Typically, if there is little wind shear, the storm will rapidly enter the dissipating stage and 'rain itself out', but, if there is sufficient change in wind speed and/or direction, the downdraft will be separated from the updraft, and the storm may become a supercell, where the mature stage can sustain itself for several hours.

Dissipating Stage

A thunderstorm in an environment with no winds to shear the storm or blow the anvil in any one direction

In the dissipation stage, the thunderstorm is dominated by the downdraft. If atmospheric conditions do not support super cellular development, this stage occurs rather quickly, approximately 20–30 minutes into the life of the thunderstorm. The downdraft will push down out of the thunderstorm, hit the ground and spread out. This phenomenon is known as a downburst. The cool air carried to the ground by the downdraft cuts off the inflow of the thunderstorm, the updraft disappears and the thunderstorm will dissipate. Thunderstorms in an atmosphere with virtually no vertical wind shear weaken as soon as they send out an outflow boundary in all directions, which then quickly cuts off its inflow of relatively warm, moist air, and kills the thunderstorm's further growth. The downdraft hitting the ground creates an outflow boundary. This can cause downbursts, a potential hazardous condition for aircraft to fly through, as a substantial change in wind speed and direction occurs, resulting in a decrease of airspeed and the subsequent reduction in lift for the aircraft. The stronger the outflow boundary is, the stronger the resultant vertical wind shear becomes.

Classification

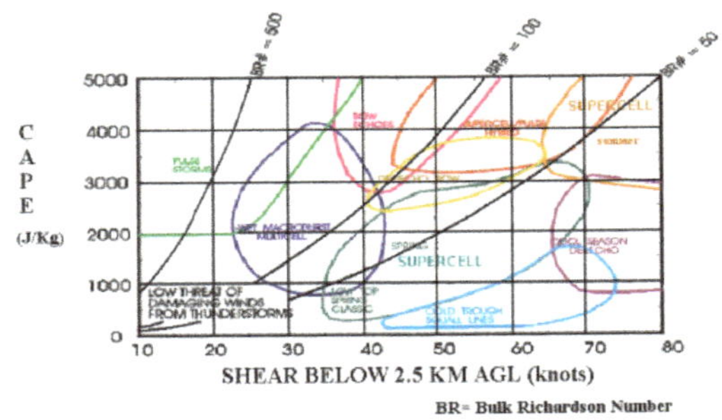

Conditions favorable for thunderstorm types and complexes

There are four main types of thunderstorms: single-cell, multi-cell, squall line (also called multi-cell line) and supercell. Which type forms depends on the instability and relative wind conditions at different layers of the atmosphere ("wind shear"). Single-cell thunderstorms form in environments of low vertical wind shear and last only 20–30 minutes. Organized thunderstorms and thunderstorm clusters/lines can have longer life cycles as they form in environments of significant vertical wind shear, which aids the development of stronger updrafts as well as various forms of severe weather. The supercell is the strongest of the thunderstorms, most commonly associated with large hail, high winds, and tornado formation.

Single-cell

A single-cell thunderstorm over Wagga Wagga.

This term technically applies to a single thunderstorm with one main updraft. Also known as air-mass thunderstorms, these are the typical summer thunderstorms in many temperate locales. They also occur in the cool unstable air that often follows the passage of a cold front from the sea during winter. Within a cluster of thunderstorms, the term "cell" refers to each separate principal updraft. Thunderstorm cells occasionally form in isolation, as the occurrence of one thunderstorm can develop an outflow boundary that sets up new thunderstorm development. Such storms are rarely severe and are a result of local atmospheric instability; hence the term "air mass thunderstorm". When such storms have a brief period of severe weather associated with them, it is known as a pulse severe storm. Pulse severe storms are poorly organized and occur randomly in time and space, making them difficult to forecast. Single-cell thunderstorms normally last 20–30 minutes.

Multi-cell Clusters

A group of thunderstorms over Brazil photographed by the Space Shuttle Challenger.

This is the most common type of thunderstorm development. *Mature thunderstorms* are found near the center of the cluster, while dissipating thunderstorms exist on their downwind side. *Multicell storms* form as clusters of storms but may then evolve into one or more squall lines. While each cell of the cluster may only last 20 minutes, the cluster itself may persist for hours at a time. They often arise from convective updrafts in or near mountain ranges and linear weather boundaries, such as strong cold fronts or troughs of low pressure. These type of storms are stronger than the single-cell storm, yet much weaker than the supercell storm. Hazards with the multicell cluster include moderate-sized hail, flash flooding, and weak tornadoes.

Multicell Lines

A squall line is an elongated line of severe thunderstorms that can form along and/or ahead of a cold front. In the early 20th century, the term was used as a synonym for cold front. The squall line contains heavy precipitation, hail, frequent lightning, strong straight line winds, and possibly tornadoes and waterspouts. Severe weather in the form of strong straight-line winds can be expected in areas where the squall line itself is in the shape of a bow echo, within the portion of the line that bows out the most. Tornadoes can be found along waves within a line echo wave pattern, or LEWP, where mesoscale low pressure areas are present. Some bow echoes in the summer are called derechos, and move quite fast through large sections of territory. On the back edge of the rain shield associated with mature squall lines, a wake low can form, which is a mesoscale low pressure area that forms behind the mesoscale high pressure system normally present under the rain canopy, which are sometimes associated with a heat burst. This kind of storm is also known as "Wind of the Stony Lake" in southern China.

Supercells

A supercell thunderstorm over Chaparral, New Mexico.

The setting sun illuminates the top of a classic anvil-shaped thunderstorm cloud in eastern Nebraska, United States.

Supercell storms are large, usually severe, quasi-steady-state storms that form in an environment where wind speed or wind direction varies with height ("wind shear"), and they have separate downdrafts and updrafts (i.e., where its associated precipitation is not falling through the updraft) with a strong, rotating updraft (a "mesocyclone"). These storms normally have such powerful updrafts that the top of the supercell storm cloud (or anvil) can break through the troposphere and reach into the lower levels of the stratosphere, and supercell storms can be 24 kilometres (15 mi) wide. Research has shown that at least 90 percent of supercells cause severe weather. These storms can produce destructive tornadoes, sometimes F3 or higher, extremely large hailstones (10 centimetres or 4 inches diameter), straight-line winds in excess of 130 km/h (81 mph), and flash floods. In fact, research has also shown that most tornadoes occur from this type of thunderstorm. Supercells are the strongest type of thunderstorm, even stronger than the most severe thunderstorms.

Severe Thunderstorms

A thunderstorm is classed as severe if winds reach at least 93 kilometres per hour (58 mph), hail is 25 millimetres (1 in) in diameter or larger, or if funnel clouds or tornadoes are reported. Although a funnel cloud or tornado indicates a severe thunderstorm, a tornado warning is issued in place of a severe thunderstorm warning. A severe thunderstorm warning is issued if a thunderstorm becomes severe, or will soon turn severe. In Canada, a rainfall rate greater than 50 millimetres (2 in) in one hour, or 75 millimetres (3 in) in three hours, is also used to indicate severe thunderstorms. Severe thunderstorms can occur from any type of storm cell. However, multicell, supercell, and squall lines represent the most common forms of thunderstorms that produce severe weather.

Mesoscale Convective Systems

MCC moving through New England: August 2, 2006 0600 UTC

A mesoscale convective system (MCS) is a complex of thunderstorms that becomes organized on a scale larger than the individual thunderstorms but smaller than extratropical cyclones, and normally persists for several hours or more. A mesoscale convective system's overall cloud and precipitation pattern may be round or linear in shape, and include weather systems such as tropical cyclones, squall lines, lake-effect snow events, polar lows, and Mesoscale Convective Complexes (MCCs), and they generally form near weather fronts. Most mesoscale convective systems develop overnight and continue their lifespan through the next day. The type that forms during the warm season over land has been noted across North America, Europe, and Asia, with a maximum in activity noted during the late afternoon and evening hours.

Forms of MCS that develop in the tropics are found in use either the Intertropical Convergence Zone or monsoon troughs, generally within the warm season between spring and fall. More intense systems form over land than over water. One exception is that of lake-effect snow bands, which form due to cold air moving across relatively warm bodies of water, and occurs from fall through spring. Polar lows are a second special class of MCS. They form at high latitudes during the cold season. Once the parent MCS dies, later thunderstorm development can occur in connection with its remnant mesoscale convective vortex (MCV). Mesoscale convective systems are important to the United States rainfall climatology over the Great Plains since they bring the region about half of their annual warm season rainfall.

Motion

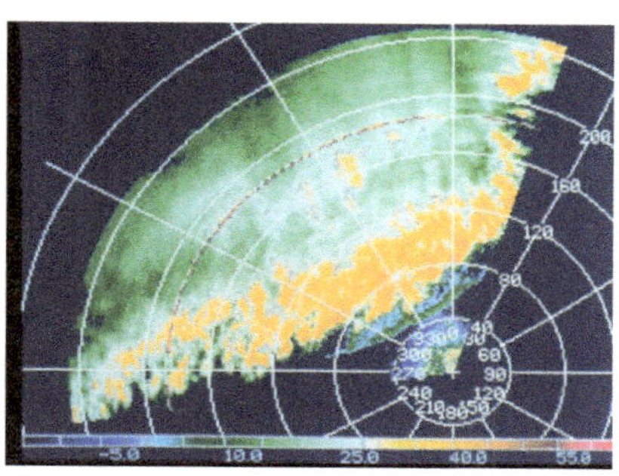

Thunderstorm line viewed in reflectivity (dBZ) on a plan position indicator radar display

The two major ways thunderstorms move are via advection of the wind and propagation along outflow boundaries towards sources of greater heat and moisture. Many thunderstorms move with the mean wind speed through the Earth's troposphere, the lowest 8 kilometres (5.0 mi) of the Earth's atmosphere. Weaker thunderstorms are steered by winds closer to the Earth's surface than stronger thunderstorms, as the weaker thunderstorms are not as tall. Organized, long-lived thunderstorm cells and complexes move at a right angle to the direction of the vertical wind shear vector. If the gust front, or leading edge of the outflow boundary, races ahead of the thunderstorm, its motion will accelerate in tandem. This is more of a factor with thunderstorms with heavy precipitation (HP) than with thunderstorms with low precipitation (LP). When thunderstorms merge, which is most likely when numerous thunderstorms exist in proximity to each other, the motion of the stronger thunderstorm normally dictates the future motion of the merged cell. The stronger the mean wind, the less likely other processes will be involved in storm motion. On weather radar, storms are tracked by using a prominent feature and tracking it from scan to scan.

Back-building Thunderstorm

A back-building thunderstorm, commonly referred to as a training thunderstorm, is a thunderstorm in which new development takes place on the upwind side (usually the west or southwest side in the Northern Hemisphere), such that the storm seems to remain stationary or propagate in a backward direction. Though the storm often appears stationary on radar, or even moving upwind, this is an illusion. The storm is really a multi-cell storm with new, more vigorous cells that form on the upwind side, replacing older cells that continue to drift downwind. When this happens, catastrophic flooding is possible. In Rapid City, South Dakota, in 1972, an unusual alignment of winds at various levels of the atmosphere combined to produce a continuously training set of cells that dropped an enormous quantity of rain upon the same area, resulting in devastating flash flooding. A similar event occurred in Boscastle, England, on 16 August 2004.

Hazards

Each year, many people are killed or seriously injured by severe thunderstorms despite the advance warning. While severe thunderstorms are most common in the spring and summer, they can occur at just about any time of the year.

Cloud-to-ground Lightning

A return stroke, cloud-to-ground lightning strike during a thunderstorm.

Cloud-to-ground lightning frequently occurs within the phenomena of thunderstorms and have numerous hazards towards landscapes and populations. One of the more significant hazards lightning can pose is the wildfires they are capable of igniting. Under a regime of low precipitation (LP) thunderstorms, where little precipitation is present, rainfall cannot prevent fires from starting when vegetation is dry as lightning produces a concentrated amount of extreme heat. Direct damage caused by lightning strikes occurs on occasion. In areas with a high frequency for cloud-to-ground lightning, like Florida, lightning causes several fatalities per year, most commonly to people working outside.

Precipitation with low potential of hydrogen levels (pH), otherwise known as acid rain, is also a frequent risk produced by lightning. Distilled water, which contains no carbon dioxide, has a neutral pH of 7. Liquids with a pH less than 7 are acidic, and those with a pH greater than 7 are basic. "Clean" or unpolluted rain has a slightly acidic pH of about 5.2, because carbon dioxide and water in the air react together to form carbonic acid, a weak acid (pH 5.6 in distilled water), but unpolluted rain also contains other chemicals. Nitric oxide present during thunderstorm phenomena, caused by the splitting of nitrogen molecules, can result in the production of acid rain, if nitric oxide forms compounds with the water molecules in precipitation, thus creating acid rain. Acid rain can damage infrastructures containing calcite or other solid chemical compounds containing carbon. In ecosystems, acid rain can dissolve plant tissues of vegetations and increase acidification process in bodies of water and in soil, resulting in deaths of marine and terrestrial organisms.

Hail

Hailstorm in Bogotá, Colombia.

Any thunderstorm that produces hail that reaches the ground is known as a hailstorm. Thunderclouds that are capable of producing hailstones are often seen obtaining green coloration. Hail is more common along mountain ranges because mountains force horizontal winds upwards (known as orographic lifting), thereby intensifying the updrafts within thunderstorms and making hail more likely. One of the more common regions for large hail is across mountainous northern India, which reported one of the highest hail-related death tolls on record in 1888. China also experiences significant hailstorms. Across Europe, Croatia experiences frequent occurrences of hail.

In North America, hail is most common in the area where Colorado, Nebraska, and Wyoming meet, known as "Hail Alley." Hail in this region occurs between the months of March and October during the afternoon and evening hours, with the bulk of the occurrences from May through September. Cheyenne, Wyoming is North America's most hail-prone city with an average of nine to ten hailstorms per season. In South America, areas prone to hail are cities like Bogotá, Colombia.

Hail can cause serious damage, notably to automobiles, aircraft, skylights, glass-roofed structures, livestock, and most commonly, farmers' crops. Hail is one of the most significant thunderstorm hazards to aircraft. When hail stones exceed 13 millimetres (1 in) in diameter, planes can be seriously damaged within seconds. The hailstones accumulating on the ground can also be hazardous to landing aircraft. Wheat, corn, soybeans, and tobacco are the most sensitive crops to hail damage. Hail is one of Canada's most costly hazards. Hailstorms have been the cause of costly and deadly events throughout history. One of the earliest recorded incidents occurred around the 9th century in Roopkund, Uttarakhand, India. The largest hailstone in terms of maximum circumference and length ever recorded in the United States fell in 2003 in Aurora, Nebraska, United States.

Tornadoes and Waterspouts

In June 2007, the town of Elie, Manitoba was struck by an F5 tornado.

A tornado is a violent, rotating column of air in contact with both the surface of the earth and a cumulonimbus cloud (otherwise known as a thundercloud) or, in rare cases, the base of a cumulus cloud. Tornadoes come in many sizes but are typically in the form of a visible condensation funnel, whose narrow end touches the earth and is often encircled by a cloud of debris and dust. Most tornadoes have wind speeds between 40 and 110 mph (64 and 177 km/h), are approximately 75 metres (246 ft) across, and travel several kilometers (a few miles) before dissipating. Some attain wind speeds of more than 300 mph (480 km/h), stretch more than 1,600 metres (1 mi) across, and stay on the ground for more than 100 kilometres (dozens of miles).

Formation of numerous waterspouts in the Great Lakes region. (North America)

The Fujita scale and the Enhanced Fujita Scale rate tornadoes by damage caused. An EF0 tornado, the weakest category, damages trees but not substantial structures. An EF5 tornado, the strongest category, rips buildings off their foundations and can deform large skyscrapers. The similar TORRO scale ranges from a T0 for extremely weak tornadoes to T11 for the most powerful known tornadoes. Doppler radar data, photogrammetry, and ground swirl patterns (cycloidal marks) may also be analyzed to determine intensity and award a rating.

A flash flood caused by a severe thunderstorm

Waterspouts have similar characteristics as tornadoes, characterized by a spiraling funnel-shaped wind current that form over bodies of water, connecting to large cumulonimbus clouds. Waterspouts are generally classified as forms of tornadoes, or more specifically, non-supercelled tornadoes that develop over large bodies of water. These spiralling columns of air are frequently developed within tropical areas close to the equator, but are less common within areas of high latitude.

Flash Flood

Flash flooding is the process where a landscape, most notably an urban environment, is subjected to rapid floods. These rapid floods occur more quickly and are more localized than seasonal river flooding or areal flooding and are frequently (though not always) associated with intense rainfall. Flash flooding can frequently occur in slow-moving thunderstorms and is usually caused by the heavy liquid precipitation that accompanies it. Flash floods are most common in densely populated urban environments, where few plants and bodies of water are present to absorb and contain the extra water. Flash flooding can be hazardous to small infrastructure, such as bridges, and weakly constructed buildings. Plants and crops in agricultural areas can be destroyed and devastated by the force of raging water. Automobiles parked within affected areas can also be displaced. Soil erosion can occur as well, exposing risks of landslide phenomena.

Downburst

Downburst winds can produce numerous hazards to landscapes experiencing thunderstorms. Downburst winds are generally very powerful, and are often mistaken for wind speeds produced by tornadoes, due to the concentrated amount of force exerted by their straight-horizontal characteristic. Downburst winds can be hazardous to unstable, incomplete, or weakly constructed infrastructures and buildings. Agricultural crops, and other plants in nearby environments can be uprooted and damaged. Aircraft engaged in takeoff or landing can crash. Automobiles can be displaced by the force exerted by downburst winds. Downburst winds are usually formed in areas when high pressure air systems of downdrafts begin to sink and displace the air masses below it, due to their higher density. When these downdrafts reach the surface, they spread out and turn into the destructive straight-horizontal winds.

Trees uprooted or displaced by the force of a downburst wind in northwest Monroe County, Wisconsin.

Thunderstorm Asthma

Thunderstorm asthma is the triggering of an asthma attack by environmental conditions directly caused by a local thunderstorm. During a thunderstorm, pollen grains can absorb moisture and then burst into much smaller fragments with these fragments being easily dispersed by wind. While larger pollen grains are usually filtered by hairs in the nose, the smaller pollen fragments are able to pass through and enter the lungs, triggering the asthma attack.

Safety Precautions

Most thunderstorms come and go fairly uneventfully; however, any thunderstorm can become severe, and all thunderstorms, by definition, present the danger of lightning. Thunderstorm preparedness and safety refers to taking steps before, during, and after a thunderstorm to minimize injury and damage.

Preparedness

Preparedness refers to precautions that should be taken before a thunderstorm. Some preparedness takes the form of general readiness (as a thunderstorm can occur at any time of the day or year). Preparing a family emergency plan, for example, can save valuable time if a storm arises quickly and unexpectedly. Preparing the home by removing dead or rotting limbs and trees, which can be blown over in high winds, can also significantly reduce the risk of property damage and personal injury.

The National Weather Service (NWS) in the United States recommends several precautions that people should take if thunderstorms are likely to occur:

- Know the names of local counties, cities, and towns, as these are how warnings are described.
- Monitor forecasts and weather conditions and know whether thunderstorms are likely in the area.

- Be alert for natural signs of an approaching storm.
- Cancel or reschedule outdoor events (to avoid being caught outdoors when a storm hits).
- Take action early so you have time to get to a safe place.
- Get inside a substantial building or hard-topped metal vehicle before threatening weather arrives.
- If you hear thunder, get to the safe place immediately.
- Avoid open areas like hilltops, fields, and beaches, and don't be or be near the tallest objects in an area when thunderstorms are occurring.
- Don't shelter under tall or isolated trees during thunderstorms.
- If in the woods, put as much distance between you and any trees during thunderstorms.
- If in a group, spread out so that you increase the chances for survivors who could come to the aid of any victims from a lightning strike.

Safety

While safety and preparedness often overlap, "thunderstorm safety" generally refers to what people should do during and after a storm. The American Red Cross recommends that people follow these precautions if a storm is imminent or in progress:

- Take action immediately upon hearing thunder. Anyone close enough to the storm to hear thunder can be struck by lightning.
- Avoid electrical appliances, including corded telephones. Cordless and wireless telephones are safe to use during a thunderstorm.
- Close and stay away from windows and doors, as glass can become a serious hazard in high wind.
- Do not bathe or shower, as plumbing conducts electricity.
- If driving, safely exit the roadway, turn on hazard lights, and park. Remain in the vehicle and avoid touching metal.

The NWS stopped recommending the "lightning crouch" in 2008 as it doesn't provide a significant level of protection and will not significantly lower the risk of being killed or injured from a nearby lightning strike.

Frequent Occurrences

Thunderstorms occur throughout the world, even in the polar regions, with the greatest frequency in tropical rainforest areas, where they may occur nearly daily. At any given time approximately 2,000 thunderstorms are occurring on Earth. Kampala and Tororo in Uganda have each been mentioned as the most thunderous places on Earth, a claim also made for Bogor on Java, Indonesia and Singapore. Other cities known for frequent storm activity include Darwin, Caracas, Manila and Mumbai. Thunderstorms are associated with the various monsoon seasons around the globe, and they populate the rainbands of tropical cyclones. In temperate regions, they are most frequent in spring and summer, although they can occur along or ahead of cold fronts at any time of year. They may also occur within a cooler air mass following the passage of a cold front over a relatively warmer body of water. Thunderstorms are rare in polar regions because of cold surface temperatures.

A mild thunderstorm over Niagara Falls, Ontario.

Some of the most powerful thunderstorms over the United States occur in the Midwest and the Southern states. These storms can produce large hail and powerful tornadoes. Thunderstorms are relatively uncommon along much of the West Coast of the United States, but they occur with greater frequency in the inland areas, particularly the Sacramento and San Joaquin Valleys of California. In spring and summer, they occur nearly daily in certain areas of the Rocky Mountains as part of the North American Monsoon regime. In the Northeast, storms take on similar characteristics and patterns as the Midwest, but with less frequency and severity. During the summer, air-mass thunderstorms are an almost daily occurrence over central and southern parts of Florida.

Types of Lightning

3-second video of a lightning strike within a thunderstorm over Island in the Sky, Canyonlands National Park, Utah

Lightning is an electrical discharge that occurs in a thunderstorm. It can be seen in the form of a bright streak (or bolt) from the sky. Lightning occurs when an electrical charge is built up within a cloud, due to static electricity generated by supercooled water droplets colliding with ice crystals near the freezing level. When a large enough charge is built up, a large discharge will occur and can be seen as lightning.

Cloud to ground lightning over Pentagon City in Arlington, Virginia

Lightning storm over Sydney

The temperature of a lightning bolt can be five times hotter than the surface of the sun. Although the lightning is extremely hot, the duration is short and 90% of strike victims survive. Contrary to the popular idea that lightning does not strike twice in the same spot, some people have been struck by lightning over three times, and skyscrapers like the Empire State Building have been struck numerous times in the same storm.

The loud bang that is heard is the super heated air around the lightning bolt expanding at the speed of sound. Because sound travels much more slowly than light the flash is seen before the bang, although both occur at the same moment.

There are several types of lightning:

- In-cloud lightning is the most common. It is lightning within a cloud and is sometimes called intra-cloud or sheet lightning.
- Cloud to ground lightning is when a bolt of lightning from a cloud strikes the ground. This form poses the greatest threat to life and property.
- Ground to cloud lightning is when a lightning bolt is induced from the ground to the cloud.
- Cloud to cloud lightning is rarely seen and is when a bolt of lightning arcs from one cloud to another.

- Ball lightning is extremely rare and has several hypothesized explanations. It is seen in the form of a 15 to 50 centimeter radius ball.
- Cloud to air lightning is when lightning from a cloud hits air of a different charge.
- Dry lightning is a misnomer that refers to a thunderstorm whose precipitation does not reach the ground.
- Heat Lightning refers to a lightning flash that is seen from the horizon that does not have accompanying thunder.
- Upper-atmospheric lightning occurs above the thunderhead.
- Clear-air lightning is used widely to describe lightning that occurs with no apparent cloud close enough to have produced it. In the U.S. and Canadian Rockies, a thunderstorm can be in an adjacent valley and not be observable, (either visually or audibly), from the valley where the lightning bolt strikes. European and Asian mountainous areas experience similar events. Also in clear areas where the storm cell is on the near horizon (within 26 km (16 mi) a strike can occur, and as the storm is so far away, the strike is referred to as *clear-air*.

Energy

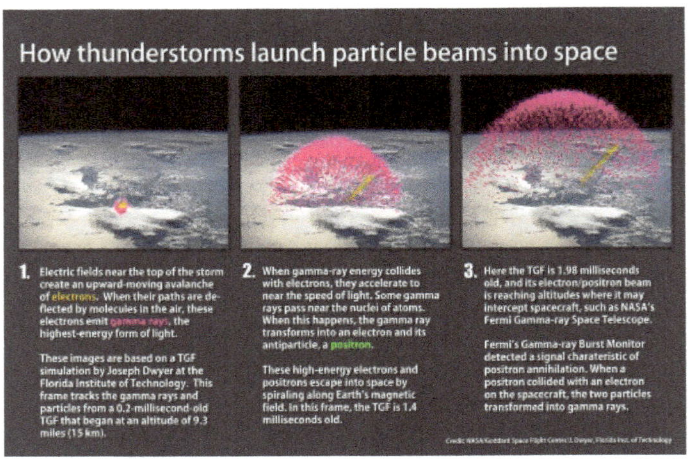

How thunderstorms launch particle beams into space

If the quantity of water that is condensed in and subsequently precipitated from a cloud is known, then the total energy of a thunderstorm can be calculated. In a typical thunderstorm, approximately 5×10^8 kg of water vapor are lifted, and the amount of energy released when this condenses is 10^{15} joules. This is on the same order of magnitude of energy released within a tropical cyclone, and more energy than that released during the atomic bomb blast at Hiroshima, Japan in 1945.

The Fermi Gamma-ray Burst Monitor results show that gamma rays and antimatter particles (positrons) can be generated in powerful thunderstorms. It is suggested that the antimatter positrons are formed in terrestrial gamma-ray flashes (TGF). TGFs are brief bursts occurring inside thunderstorms and associated with lightning. The streams of positrons and electrons collide higher in the atmosphere to generate more gamma rays. About 500 TGFs may occur every day worldwide, but mostly go undetected.

Studies

In more contemporary times, thunderstorms have taken on the role of a scientific curiosity. Every spring, storm chasers head to the Great Plains of the United States and the Canadian Prairies to explore the scientific aspects of storms and tornadoes through use of videotaping. Radio pulses produced by cosmic rays are being used to study how electric charges develop within thunderstorms. More organized meteorological projects such as VORTEX2 use an array of sensors, such as the Doppler on Wheels, vehicles with mounted automated weather stations, weather balloons, and unmanned aircraft to investigate thunderstorms expected to produce severe weather. Lightning is detected remotely using sensors that detect cloud-to-ground lightning strokes with 95 percent accuracy in detection and within 250 metres (820 ft) of their point of origin.

Mythology and Religion

Thunderstorms strongly influenced many early civilizations. Greeks believed that they were battles waged by Zeus, who hurled lightning bolts forged by Hephaestus. Some American Indian tribes associated thunderstorms with the Thunderbird, who they believed was a servant of the Great Spirit. The Norse considered thunderstorms to occur when Thor went to fight Jötnar, with the thunder and lightning being the effect of his strikes with the hammer Mjölnir. Hinduism recognizes Indra as the god of rain and thunderstorms. Christian doctrine accepted the ideas of Aristotle's original work, called *Meteorologica*, that winds were caused by exhalations from the Earth and that fierce storms were the work of God. These ideas were still within the mainstream as late as the 18th century.

Martin Luther was out walking when a thunderstorm began, causing him to pray to God for being saved and promising to become a monk.

Outside of Earth

The clouds of Venus are capable of producing lightning much like the clouds on Earth. The lightning rate is at least half of that on Earth. A thin layer of water clouds appears to underlie the ammonia layer within Jupiter's atmosphere, where thunderstorms evidenced by flashes of lightning have been detected. (Water is a polar molecule that can carry a charge, so it is capable of creating the charge separation needed to produce lightning.) These electrical discharges can be up to a thousand times as powerful as lightning on the Earth. The water clouds can form thunderstorms driven by the heat rising from the interior.

Hail

Hail is a form of solid precipitation. It is distinct from American sleet (called ice pellets outside of the United States), though the two are often confused. It consists of balls or irregular lumps of ice, each of which is called a hailstone. Sleet (ice pellets) falls generally in cold weather while hail growth is greatly inhibited during cold surface temperatures.

Unlike graupel, which is made of rime, and ice pellets, which are smaller and translucent, hailstones consist mostly of water ice and measure between 5 millimetres (0.2 in) and 15 centimetres (6 in) in diameter. The METAR reporting code for hail 5 mm (0.20 in) or greater is GR, while smaller hailstones and graupel are coded GS.

Hail is possible within most thunderstorms as it is produced by cumulonimbi, and within 2 nautical miles (3.7 km) of the parent storm. Hail formation requires environments of strong, upward motion of air with the parent thunderstorm (similar to tornadoes) and lowered heights of the freezing level. In the mid-latitudes, hail forms near the interiors of continents, while in the tropics, it tends to be confined to high elevations.

There are methods available to detect hail-producing thunderstorms using weather satellites and weather radar imagery. Hailstones generally fall at higher speeds as they grow in size, though complicating factors such as melting, friction with air, wind, and interaction with rain and other hailstones can slow their descent through Earth's atmosphere. Severe weather warnings are issued for hail when the stones reach a damaging size, as it can cause serious damage to human-made structures and, most commonly, farmers' crops.

Definition

Any thunderstorm which produces hail that reaches the ground is known as a hailstorm. Hail has a diameter of 5 millimetres (0.20 in) or more. Hailstones can grow to 15 centimetres (6 in) and weigh more than 0.5 kilograms (1.1 lb).

Unlike ice pellets, hailstones are layered and can be irregular and clumped together. Hail is composed of transparent ice or alternating layers of transparent and translucent ice at least 1 millimetre (0.039 in) thick, which are deposited upon the hailstone as it travels through the cloud, suspended aloft by air with strong upward motion until its weight overcomes the updraft and falls to the ground. Although the diameter of hail is varied, in the United States, the average observation of damaging hail is between 2.5 cm (1 in) and golf ball-sized (1.75 in).

Stones larger than 2 cm (0.80 in) are usually considered large enough to cause damage. The Meteorological Service of Canada issues severe thunderstorm warnings when hail that size or above is expected. The US National Weather Service has a 2.5 cm (1 in) or greater in diameter threshold, effective January 2010, an increase over the previous threshold of ¾-inch hail. Other countries have different thresholds according local sensitivity to hail; for instance grape growing areas could be adversely impacted by smaller hailstones. Hailstones can be very large or very small, depending on how strong the updraft is: weaker hailstorms produce smaller hailstones than stronger hailstorms (such as supercells).

Formation

Hail forms in strong thunderstorm clouds, particularly those with intense updrafts, high liquid water content, great vertical extent, large water droplets, and where a good portion of the cloud layer is below freezing 0 °C (32 °F). These types of strong updrafts can also indicate the presence of a tornado. The growth rate is maximized where air is near a temperature of −13 °C (9 °F).

Layer Nature of the Hailstones

Like other precipitation in cumulonimbus clouds, hail begins as water droplets. As the droplets rise and the temperature goes below freezing, they become supercooled water and will freeze on contact with condensation nuclei. A cross-section through a large hailstone shows an onion-like structure. This means the hailstone is made of thick and translucent layers, alternating with layers that are thin, white and opaque. Former theory suggested that hailstones were subjected to multiple descents and ascents, falling into a zone of humidity and refreezing as they were uplifted. This up and down motion was thought to be responsible for the successive layers of the hailstone. New research, based on theory as well as field study, has shown this is not necessarily true.

Hail shaft

Severe thunderstorms containing hail can exhibit a characteristic green coloration

The storm's updraft, with upwardly directed wind speeds as high as 110 miles per hour (180 km/h), blows the forming hailstones up the cloud. As the hailstone ascends it passes into areas of the cloud where the concentration of humidity and supercooled water droplets varies. The hailstone's growth rate changes depending on the variation in humidity and supercooled water droplets that it encounters. The accretion rate of these water droplets is another factor in the hailstone's growth. When the hailstone moves into an area with a high concentration of water droplets, it captures the latter and acquires a translucent layer. Should the hailstone move into an area where mostly water vapour is available, it acquires a layer of opaque white ice.

Furthermore, the hailstone's speed depends on its position in the cloud's updraft and its mass. This determines the varying thicknesses of the layers of the hailstone. The accretion rate of supercooled water droplets onto the hailstone depends on the relative velocities between these water droplets and the hailstone itself. This means that generally the larger hailstones will form some distance from the stronger updraft where they can pass more time growing. As the hailstone grows it releases latent heat, which keeps its exterior in a liquid phase. Because it undergoes 'wet growth', the outer layer is *sticky* (i.e. more adhesive), so a single hailstone may grow by collision with other smaller hailstones, forming a larger entity with an irregular shape.

The hailstone will keep rising in the thunderstorm until its mass can no longer be supported by the updraft. This may take at least 30 minutes based on the force of the updrafts in the hail-producing thunderstorm, whose top is usually greater than 10 km high. It then falls toward the ground while continuing to grow, based on the same processes, until it leaves the cloud. It will later begin to melt as it passes into air above freezing temperature.

Thus, a unique trajectory in the thunderstorm is sufficient to explain the layer-like structure of the hailstone. The only case in which multiple trajectories can be discussed is in a multicellular thunderstorm, where the hailstone may be ejected from the top of the "mother" cell and captured in the updraft of a more intense "daughter" cell. This, however, is an exceptional case.

Factors Favoring Hail

Hail is most common within continental interiors of the mid-latitudes, as hail formation is considerably more likely when the freezing level is below the altitude of 11,000 feet (3,400 m). Movement of dry air into strong thunderstorms over continents can increase the frequency of hail by promoting evaporational cooling which lowers the freezing level of thunderstorm clouds giving hail a larger volume to grow in. Accordingly, hail is less common in the tropics despite a much higher frequency of thunderstorms than in the mid-latitudes because the atmosphere over the tropics tends to be warmer over a much greater altitude. Hail in the tropics occurs mainly at higher elevations.

Hail growth becomes vanishingly small when air temperatures fall below −30 °C (−22 °F) as supercooled water droplets become rare at these temperatures. Around thunderstorms, hail is most likely within the cloud at elevations above 20,000 feet (6,100 m). Between 10,000 feet (3,000 m) and 20,000 feet (6,100 m), 60 percent of hail is still within the thunderstorm, though 40 percent now lies within the clear air under the anvil. Below 10,000 feet (3,000 m), hail is equally distributed in and around a thunderstorm to a distance of 2 nautical miles (3.7 km).

Climatology

Hail occurs most frequently within continental interiors at mid-latitudes and is less common in the tropics, despite a much higher frequency of thunderstorms than in the mid-latitudes. Hail is also much more common along mountain ranges because mountains force horizontal winds upwards (known as orographic lifting), thereby intensifying the updrafts within thunderstorms and making hail more likely. The higher elevations also result in there being less time available for hail to melt before reaching the ground. One of the more common regions for large hail is across mountainous northern India, which reported one of the highest hail-related death tolls on record in 1888. China also experiences significant hailstorms. Central Europe and southern Australia also experience a lot of hailstorms. Popular regions for hailstorms are southern and western Germany, northern and eastern France and southern and eastern Benelux. In south-eastern Europe, Croatia and Serbia experience frequent occurrences of hail.

In North America, hail is most common in the area where Colorado, Nebraska, and Wyoming meet, known as "Hail Alley". Hail in this region occurs between the months of March and October during the afternoon and evening hours, with the bulk of the occurrences from May through September. Cheyenne, Wyoming is North America's most hail-prone city with an average of nine to ten hailstorms per season.

Short-term Detection

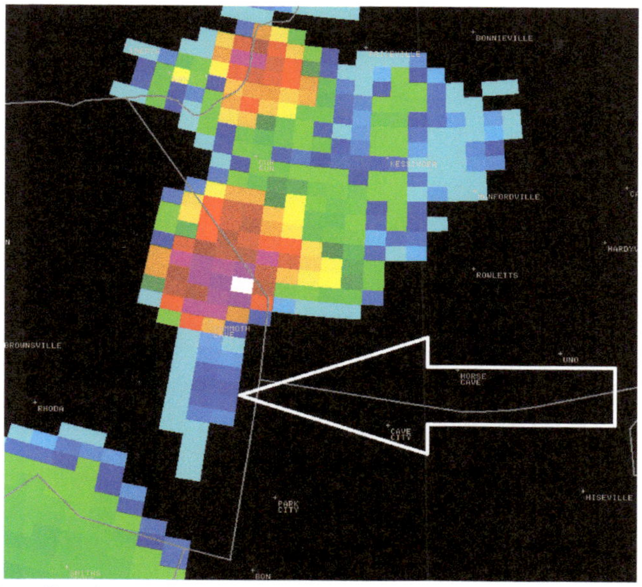

Example of a three body spike: the weak triangular echoes (pointed by the arrow) behind the red and white thunderstorm core are related to hail inside the storm.

Weather radar is a very useful tool to detect the presence of hail-producing thunderstorms. However, radar data has to be complemented by a knowledge of current atmospheric conditions which can allow one to determine if the current atmosphere is conducive to hail development.

Modern radar scans many angles around the site. Reflectivity values at multiple angles above ground level in a storm are proportional to the precipitation rate at those levels. Summing reflectivities in the Vertically Integrated Liquid or VIL, gives the liquid water content in the cloud. Research shows that hail development in the upper levels of the storm is related to the evolution of VIL. VIL divided by the vertical extent of the storm, called VIL density, has a relationship with hail size, although this varies with atmospheric conditions and therefore is not highly accurate. Traditionally, hail size and probability can be estimated from radar data by computer using algorithms based on this research. Some algorithms include the height of the freezing level to estimate the melting of the hailstone and what would be left on the ground.

Certain patterns of reflectivity are important clues for the meteorologist as well. The three body scatter spike is an example. This is the result of energy from the radar hitting hail and being deflected to the ground, where they deflect back to the hail and then to the radar. The energy took more time to go from the hail to the ground and back, as opposed to the energy that went direct from the hail to the radar, and the echo is further away from the radar than the actual location of the hail on the same radial path, forming a cone of weaker reflectivities.

More recently, the polarization properties of weather radar returns have been analyzed to differentiate between hail and heavy rain. The use of differential reflectivity (Z_{dr}), in combination with horizontal reflectivity (Z_{dr}) has led to a variety of hail classification algorithms. Visible satellite imagery is beginning to be used to detect hail, but false alarm rates remain high using this method.

Size and Terminal Velocity

Hailstones ranging in size from few millimetres to over a centimetre in diameter.

Large hailstone with concentric rings

The size of hailstones is best determined by measuring their diameter with a ruler. In the absence of a ruler, hailstone size is often visually estimated by comparing its size to that of known objects, such as coins. Using the objects such as hen's eggs, peas, and marbles for comparing hailstone sizes is imprecise, due to their varied dimensions. The UK organisation, TORRO, also scales for both hailstones and hailstorms. When observed at an airport, METAR code is used within a surface weather observation which relates to the size of the hailstone. Within METAR code, GR is used to indicate larger hail, of a diameter of at least 0.25 inches (6.4 mm). GR is derived from the French word grêle. Smaller-sized hail, as well as snow pellets, use the coding of GS, which is short for the French word grésil.

The largest recorded hailstone in the United States by diameter 7.9 inches (20 cm) and weight 1.94 pounds (0.88 kg). The hailstone fell in Vivian, South Dakota on July 23, 2010.

Hail Records

Megacryometeors, large rocks of ice that are not associated with thunderstorms, are not officially recognized by the World Meteorological Organization as "hail," which are aggregations of ice associated with thunderstorms, and therefore records of extreme characteristics of megacryometeors are not given as hail records.

- Heaviest: 1.02 kg (2.25 lb); Gopalganj District, Bangladesh, 14 April 1986.
- Largest diameter officially measured: 7.9 inches (20 cm) diameter, 18.622 inches (47.3 cm) circumference; Vivian, South Dakota, 23 July 2010.
- Largest circumference officially measured: 18.74 inches (47.6 cm) circumference, 7.0 inches (17.8 cm) diameter; Aurora, Nebraska, 22 June 2003.
- Kericho hills in Kenya experiences hailstorms, on average, around 132 days annually. Kericho hills is one of the main tea-producing regions in the world, and it is likely that dust particles, created by the activities of tea-pickers, "act as nucleation points in the atmosphere around which hailstorms form."

Terminal velocity of hail, or the speed at which hail is falling when it strikes the ground, varies. It is estimated that a hailstone of 1 centimetre (0.39 in) in diameter falls at a rate of 9 metres per second (20 mph), while stones the size of 8 centimetres (3.1 in) in diameter fall at a rate of 48 metres per second (110 mph). Hailstone velocity is dependent on the size of the stone, friction with air it is falling through, the motion of wind it is falling through, collisions with raindrops or other hailstones, and melting as the stones fall through a warmer atmosphere. As hail stones are not perfect spheres it is difficult to calculate their speed accurately.

Hazards

Hail can cause serious damage, notably to automobiles, aircraft, skylights, glass-roofed structures, livestock, and most commonly, farmers' crops. Hail damage to roofs often goes unnoticed until further structural damage is seen, such as leaks or cracks. It is hardest to recognize hail damage on shingled roofs and flat roofs, but all roofs have their own hail damage detection problems. Metal roofs are fairly resistant to hail damage, but may accumulate cosmetic damage in the form of dents and damaged coatings.

Hail is one of the most significant thunderstorm hazards to aircraft. When hailstones exceed 0.5 inches (13 mm) in diameter, planes can be seriously damaged within seconds. The hailstones accumulating on the ground can also be hazardous to landing aircraft. Hail is also a common nuisance to drivers of automobiles, severely denting the vehicle and cracking or even shattering windshields and windows. Wheat, corn, soybeans, and tobacco are the most sensitive crops to hail damage. Hail is one of Canada's most expensive hazards. Rarely, massive hailstones have been known to cause concussions or fatal head trauma. Hailstorms have been the cause of costly and deadly events throughout history. One of the earliest known incidents occurred around the 9th century in Roopkund, Uttarakhand, India, where 200 to 600 nomads seem to have died of injuries from hail the size of cricket balls. The largest hailstone in terms of diameter and weight ever recorded in the United States fell on July 23, 2010 in Vivian, South Dakota; it measured 7.9 inches (20 cm) in diameter and 18.62 inches (47.3 cm) in circumference, weighing in at 1.94 pounds (0.88 kg). This broke the previous record for diameter set by a hailstone 7 inches diameter and 18.74 inches circumference (still the greatest *circumference* hailstone) which fell in Aurora, Nebraska in the United States on June 22, 2003, as well as the record for weight, set by a hailstone of 1.67 pounds (0.76 kg) that fell in Coffeyville, Kansas in 1970.

Accumulations

Accumulated hail after a storm in Charlton, Massachusetts

Narrow zones where hail accumulates on the ground in association with thunderstorm activity are known as hail streaks or hail swaths, which can be detectable by satellite after the storms pass by. Hailstorms normally last from a few minutes up to 15 minutes in duration. Accumulating hail storms can blanket the ground with over 2 inches (5.1 cm) of hail, cause thousands to lose power, and bring down many trees. Flash flooding and mudslides within areas of steep terrain can be a concern with accumulating hail.

Depths of up to 18 in (0.46 m) have been reported. A landscape covered in accumulated hail generally resembles one covered in accumulated snow and any significant accumulation of hail has the same restrictive effects as snow accumulation, albeit over a smaller area, on transport and infrastructure. Accumulated hail can also cause flooding by blocking drains, and hail can be carried in the floodwater, turning into a snow-like slush which is deposited at lower elevations.

On somewhat rare occasions, a thunderstorm can become stationary or nearly so while prolifically producing hail and significant depths of accumulation do occur; this tends to happen in mountainous areas, such as the July 29, 2010 case of a foot of hail accumulation in Boulder County, Colorado. On June 5, 2015, hail up to four feet deep fell on one city block in Denver, Colorado. The hailstones, described as between the size of bumble bees and ping pong balls, were accompanied by rain and high winds. The hail fell in only the one area, leaving the surrounding area untouched. It fell for one and a half hours between 10 p.m. and 11:30 p.m. A meteorologist for the National Weather Service in Boulder said, “It’s a very interesting phenomenon. We saw the storm stall. It produced copious amounts of hail in one small area. It’s a meteorological thing.” Tractors used to clear the area filled more than 30 dump-truck loads of hail.

Research focused on four individual days that accumulated more than 5.9 inches (15 cm) of hail in 30 minutes on the Colorado front range has shown that these events share similar patterns in observed synoptic weather, radar, and lightning characteristics, suggesting the possibility of predicting these events prior to their occurrence. A fundamental problem in continuing research in this area is that, unlike hail diameter, hail depth is not commonly reported. The lack of data leaves researchers and forecasters in the dark when trying to verify operational methods. A cooperative effort between the University of Colorado and the National Weather Service, for the summer of 2016, is in progress. The joint project's goal is to enlist the help of the general public to develop a database of hail accumulation depths.

Suppression and Prevention

During the Middle Ages, people in Europe used to ring church bells and fire cannons to try to prevent hail, and the subsequent damage to crops. Updated versions of this approach are available as modern hail cannons. Cloud seeding after World War II was done to eliminate the hail threat, particularly across the Soviet Union – where it was claimed a 70 to 98 percent reduction in crop damage from hail storms was achieved by deploying silver iodide in clouds using rockets and artillery shells. Hail suppression programs have been undertaken by 15 countries between 1965 and 2005.

References

- D. Alexander (2002). Principles of Emergency planning and Management. Harpended: Terra publishing. ISBN 1-903544-10-6.
- B. Wisner; P. Blaikie; T. Cannon & I. Davis (2004). At Risk - Natural hazards, people's vulnerability and disasters. Wiltshire: Routledge. ISBN 0-415-25216-4.
- Ohnaka, M. (2013). The Physics of Rock Failure and Earthquakes. Cambridge University Press. p. 148. ISBN 9781107355330.
- Van Riper, A. Bowdoin (2002). Science in popular culture: a reference guide. Westport: Greenwood Press. p. 60. ISBN 0-313-31822-0.
- Thomas, David; Goudie, Andrew, eds. (2009). The Dictionary of Physical Geography (3rd ed.). Chichester: John Wiley & Sons. p. 440. ISBN 1444313169.
- Waltham, Tony; Bell, Fred; Culshaw, Martin (2005). Sinkholes and subsidence: karst and cavernous rocks in engineering and construction (1st ed.). Berlin [u.a.]: Springer [u.a.] p. 64. ISBN 3540207252.
- Abe K. (1995). Estimate of Tsunami Run-up Heights from Earthquake Magnitudes. Tsunami: progress in prediction, disaster prevention, and warning. ISBN 978-0-7923-3483-5. Retrieved 2009-10-18.
- Erik A. Rasmussen; John Turner (2003). Polar lows: mesoscale weather systems in the polar regions. Cambridge University Press. p. 224. ISBN 978-0-521-62430-5. Retrieved 2011-01-27.
- E. A. Rasmussen; J. Turner (2003). Polar Lows: Mesoscale Weather Systems in the Polar Regions. Cambridge University Press. p. 612. ISBN 978-0-521-62430-5.
- Robert Penrose Pearce (2002). Meteorology at the Millennium. Academic Press. p. 66. ISBN 978-0-12-548035-2. Retrieved 2009-01-02.
- Roland Paepe; Rhodes Whitmore Fairbridge; Saskia Jelgersma (1990). Greenhouse Effect, Sea Level and Drought. Springer Science & Business Media. p. 22. ISBN 0792310179. Retrieved 2015-02-18.
- Joseph S. D'Aleo; Pamela G. Grube (2002). The Oryx Resource Guide to El Niño and La Niña. Greenwood Publishing Group. pp. 48–49. ISBN 1573563781. Retrieved 2015-02-18.

- Bin Wang (2006-01-13). The Asian Monsoon. Springer Science & Business Media. p. 206. ISBN 3540406107. Retrieved 2008-05-03.
- K.E.K. Neuendorf; J.P. Mehl, Jr.; J.A. Jackson (2005). Glossary of Geology. Springer-Verlag, New York. p. 779. ISBN 3-540-27951-2.
- Arthur Getis; Judith Getis and Jerome D. Fellmann (2000). Introduction to Geography, Seventh Edition. McGraw-Hill. p. 99. ISBN 0-697-38506-X.
- Chris C. Mooney (2007). Storm world: hurricanes, politics, and the battle over global warming. Houghton Mifflin Harcourt. p. 20. ISBN 978-0-15-101287-9. Retrieved 2009-08-31.
- Michael H. Mogil (2007). Extreme Weather. New York: Black Dog & Leventhal Publisher. pp. 210–211. ISBN 978-1-57912-743-5.
- Glossary of Meteorology (2009). Line echo wave pattern. American Meteorological Society. ISBN 1-878220-34-9. Retrieved 2009-05-03.
- Michael Garstang; David Roy Fitzjarrald (1999). Observations of surface to atmosphere interactions in the tropics. Oxford University Press US. pp. 40–41. ISBN 978-0-19-511270-2.
- E. A. Rasmussen & J. Turner (2003). Polar Lows: Mesoscale Weather Systems in the Polar Regions. Cambridge University Press. p. 612. ISBN 978-0-521-62430-5.
- Damon P. Coppola (2007). Introduction to international disaster management. Butterworth-Heinemann. p. 62. ISBN 978-0-7506-7982-4.
- Thomas E. Downing; Alexander A. Olsthoorn; Richard S. J. Tol (1999). Climate, change and risk. Routledge. pp. 41–43. ISBN 978-0-415-17031-4. Retrieved 2009-07-16.
- Damon P. Coppola (2007). Introduction to international disaster management. Butterworth-Heinemann. p. 62. ISBN 978-0-7506-7982-4.

3

Various Man-made Disasters

Man-made disasters are disasters which are the consequences of technological hazards. The examples of man-made disasters are terrorism, sabotage, ecotage, genocides, riots and massacres. Wars and deliberate attacks are also a part of this category of disasters. This section discusses man-made disasters in a critical manner providing key analysis to the subject matter.

Anthropogenic Hazard

Anthropogenic hazards are those hazards caused directly or indirectly by human action or inaction. They can be contrasted with natural hazards. Anthropogenic hazards may adversely affect humans, other organisms and biomes and eco-systems. The frequency and severity of hazards are key elements in some risk analysis methodologies. Hazards may also be described in relation to the impact that they have. Environmental hazards may be very different from human hazards. A hazard only exists if there is a pathway to exposure. As an example the center of the earth consists of molten material at very high temperatures which would be a severe hazard if contact was made with the core. However, there is no feasible way of making contact with the core, therefore the center of the earth currently poses no hazard.

Societal Hazards

Societal hazards occur when the behavior of individuals or groups of individuals give rise to a reasonable apprehension of injury or death.

Criminality

Behavior which puts others at risk of injury or death is universally regarded as criminal and is a breach of the law for which the appropriate legal authority may impose some form of penalty, such as imprisonment, a fine, or even execution. Understanding what makes individuals act in a way that puts others at risk has been the subject of much research in many developed countries. Mitigating the hazard of criminality is very dependent on time and place with some areas and times of day posing a greater risk than others.

Civil Disorder

Civil disorder is a broad term that is typically used by law enforcement to describe forms of disturbance when many people are involved and are set upon a common aim. Civil disorder has many causes, including large-scale criminal conspiracy, socio-economic factors (unemployment, poverty), hostility between racial and ethnic groups and outrage over perceived moral and legal transgressions. Examples of well-known civil disorders and riots are the Poll Tax Riots in the United Kingdom in 1990; the 1992 Los Angeles riots in which 53 people died;

the 2008 Greek riots after a 15-year-old boy was fatally shot by police; and the 2010 Thai political protests in Bangkok during which 91 people died. Such behavior is only hazardous for those directly involved as participants or those controlling the disturbance or those indirectly involved as passers-by or shopkeepers for example. For the great majority, staying out of the way of the disturbance eliminates the hazard.

Terrorism

The common definition of terrorism is the use or threatened use of violence for the purpose of creating fear in order to achieve a political, religious, or ideological goal. Targets of terrorist acts can be anyone, including private citizens, government officials, military personnel, law enforcement officers, firefighters, or people serving in the interests of governments.

Definitions of terrorism may also vary geographically. In Australia, the Security Legislation Amendment (Terrorism) Act 2002, defines terrorism as "an action to advance a political, religious or ideological cause and with the intention of coercing the government or intimidating the public", while the United States Department of State operationally describes it as "premeditated, politically-motivated violence perpetrated against noncombatant targets by sub national groups or clandestine agents, usually intended to influence an audience".

War

War is a conflict between relatively large groups of people, which involves physical force inflicted by the use of weapons. Warfare has destroyed entire cultures, countries, economies and inflicted great suffering on humanity. Other terms for war can include armed conflict, hostilities, and police action. Acts of war are normally excluded from insurance contracts and sometimes from disaster planning.

Industrial Hazards

Industrial accidents resulting in releases of hazardous materials usually occur in a commercial context, such as mining accidents. They often have an environmental impact but also can be hazardous for people living in proximity. The Bhopal disaster saw the release of Dioxins into the neighbouring environment seriously affecting large numbers of people. It is probably the world's worst industrial accident to date.

Engineering Hazards

Engineering hazards occur when structures used by people fail or the materials used in their construction prove to be hazardous. This history of construction has many examples of hazards associated with structures including bridge failures such as the Tay Bridge disaster caused by under-design, the Silver Bridge collapse caused by corrosion attack or the original Tacoma Narrows Bridge caused by aerodynamic flutter of the deck. Failure of dams was not infrequent during the Victorian era, such as the Dale Dyke dam failure in Sheffield, England in 1864, causing the Great Sheffield Flood, which killed at least 240 people. In 1889, the failure of the South Fork Dam on the Little Conemaugh River near Johnstown, Pennsylvania, produced the Johnstown Flood, which killed over 2,200. Other failures include balcony collapses, aerial walkway collapses such as the Hyatt Regency walkway collapse in Kansas City in 1981, and building collapses such as that of the World Trade Center in New York City in 2001 during the September 11 attacks.

Waste Disposal

In developed economies, many hazardous materials are put in the domestic and commercial waste stream. In part this is because modern technological living uses many hazardous materials which, when they are in use, are safely encapsulated packaged to avoid any exposure route. In the waste stream, the packaging or encapsulation breaks or degrades and there is an exposure route to hazardous materials for people working in the waste disposal industry, those living around sites used for waste disposal and the general environment surrounding such sites.

Power Outage

A power outage is an interruption of normal sources of electrical power. Short-term power outages (up to a few hours) are common and have minor adverse effect, since most businesses and health facilities are prepared to deal with them. Extended power outages, however, can disrupt personal and business activities as well as medical and rescue services, leading to business losses and medical emergencies. Extended loss of power can lead to civil disorder, as in the New York City blackout of 1977. Only very rarely do power outages escalate to disaster proportions, however, they often accompany other types of disasters, such as hurricanes and floods, which hampers relief efforts.

Electromagnetic pulses and voltage spikes from whatever cause can also damage electricity infrastructure and electrical devices.

Recent notable power outages include the 2005 Java–Bali Blackout which affected 100 million people, 2012 India blackouts which affected 600 million and the 2009 Brazil and Paraguay blackout which affected 60 million people.

Fire

An active flame front of the Zaca Fire

Bush fires, forest fires, and mine fires are generally started by lightning, but also by human negligence or arson. They can burn thousands of square kilometers. If a fire intensifies enough to produce its own winds and "weather", it will form into a firestorm. A good example of a mine fire is the one near Centralia, Pennsylvania. Started in 1962, it ruined the town and continues to burn today. Some of the biggest city-related fires are The Great Chicago Fire, The Peshtigo Fire (both of 1871) and the Great Fire of London in 1666.

Casualties resulting from fires, regardless of their source or initial cause, can be aggravated by inadequate emergency preparedness. Such hazards as a lack of accessible emergency exits, poorly marked escape routes, or improperly maintained fire extinguishers or sprinkler systems may result in many more deaths and injuries than might occur with such protections.

A building damaged by arson

Arson is the setting a fire with intent to cause damage. The definition of arson was originally limited to setting fire to buildings, but was later expanded to include other objects, such as bridges, vehicles, and private property. Some human-induced fires are accidental: failing machinery such as a kitchen stove is a major cause of accidental fires.

Hazardous Materials

Organohalogens

Organohalogens are a family of synthetic organic molecules which all contain atoms of one of the halogens. Such materials include PCBs, Dioxins, DDT, Freon and many others. Although considered harmless when first produced, many of these compounds are now known to have profound physiological effects on many organisms including man. Many are also fat soluble and become concentrated through the food chain.

Toxic Metals

Many metals and their salts can exhibit toxicity to humans and many other organisms. Such metals include, Lead, Cadmium, Copper, Silver, Mercury and many of the transuranic metals.

Radioactive Materials

Radioactive materials produce ionizing radiation which may be very harmful to living organisms. Damage from even a short exposure to radioactivity may have long term adverse health consequences.

Chernobyl nuclear power plant

Exposure may occur from nuclear fallout when nuclear weapons are detonated or nuclear containment systems are compromised. During World War II, the United States Army Air Forces dropped atomic bombs on the Japanese cities of Hiroshima and Nagasaki, leading to extensive contamination of food, land, and water. In the Soviet Union, the Mayak industrial complex (otherwise known as Chelyabinsk-40 or Chelyabinsk-65) exploded in 1957. The Kyshtym disaster was kept secret for several decades. It is the third most serious nuclear accident ever recorded. At least 22 villages were exposed to radiation and resulted in at least 10,000 displaced persons. In 1992 the former soviet union officially acknowledge the accident. Other Soviet republics of Ukraine and Belarus suffered also when a reactor at the Chernobyl nuclear power plant had a meltdown in 1986. To this day, several small towns and the city of Chernobyl remain abandoned and uninhabitable due to fallout.

The Hanford Site is a decommissioned nuclear production complex that produced plutonium for most of the 60,000 weapons in the U.S. nuclear arsenal. There are environmental concerns about radioactivity released from Hanford.

A number of military accidents involving nuclear weapons have also resulted in radioactive contamination, for example the 1966 Palomares B-52 crash and the 1968 Thule Air Base B-52 crash.

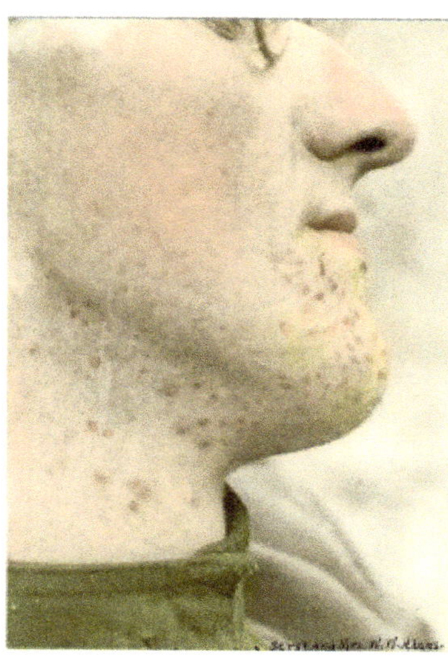

Dermatitis (burn) of chin from vapors of mustard gas

CBRNs

CBRN is a catch-all acronym for chemical, biological, radiological, and nuclear. The term is used to describe a non-conventional terror threat that, if used by a nation, would be considered use of a weapon of mass destruction. This term is used primarily in the United Kingdom. Planning for the possibility of a CBRN event may be appropriate for certain high-risk or high-value facilities and governments. Examples include Saddam Hussein's Halabja poison gas attack, the Sarin gas attack on the Tokyo subway and the preceding test runs in Matsumoto, Japan 100 kilometers outside of Tokyo, and Lord Amherst giving smallpox laden blankets to Native Americans.

Transportation

Aviation

The ditching of US Airways Flight 1549 was a well-publicised incident in which all on board survived

An aviation incident is an occurrence other than an accident, associated with the operation of an aircraft, which affects or could affect the safety of operations, passengers, or pilots. The category of the vehicle can range from a helicopter, an airliner, or a space shuttle.

Rail

Granville-Paris Express wreck at Gare Montparnasse on 22 October 1895

The special hazards of traveling by rail include the possibility of a train crash which can result in substantial loss of life. Incidents involving freight traffic generally pose a greater hazardous risk to the environment. Less common hazards include geophysical hazards such as tsunami such as that which struck in 2004 in Sri Lanka when 1,700 people died in the Sri Lanka tsunami-rail disaster.

Road

Traffic collisions are the leading cause of death, and road-based pollution creates a substantial health hazard, especially in major conurbations.

Space

Disintegration of the Space Shuttle Challenger

Space travel presents significant hazards, mostly to the direct participants (astronauts or cosmonauts and ground support personnel), but also carry the potential of disaster to the public at large. Accidents related to space travel have killed 22 astronauts and cosmonauts, and a larger number of people on the ground.

Accidents can occur on the ground during launch, preparation, or in flight, due to equipment malfunction or the naturally hostile environment of space itself. An additional risk is posed by (unmanned) low-orbiting satellites whose orbits eventually decay due to friction with the extremely thin atmosphere. If they are large enough, massive pieces traveling at great speed can fall to the Earth before burning up, with the potential to do damage.

One of the worst human-piloted space accidents involved the Space Shuttle *Challenger* which disintegrated in 1986, claiming all seven lives on board. The shuttle disintegrated 73 seconds after taking off from the launch pad in Cape Canaveral, Florida.

Another example is the Space Shuttle *Columbia*, which disintegrated during a landing attempt over Texas in 2003, with a loss of all seven astronauts on board. The debris field extended from New Mexico to Mississippi.

Sea travel

The capsized cruise ship Costa Concordia with a large rock lodged in the crushed hull of the ship

Ships can sink, capsize or crash in disasters. Perhaps the most infamous sinking was that of the Titanic which hit an iceberg and sank, resulting in one of the worst maritime disasters in history. Other notable incidents include the capsizing of the Costa Concordia, which killed at least 32 people; and is the largest passenger ship to sink, and the sinking of the MV Doña Paz, which claimed the lives of up to 4,375 people in the worst peacetime maritime disaster in history.

Environmental hazards

Environmental hazards are those hazards where the effects are seen in biomes or ecosystems rather than directly on living organisms. Well known examples include oil spills, water pollution, slash and burn de-forestation etc.

- Environmental disaster
- Existential threat
- List of wars and anthropogenic disasters by death toll
- Survivalism
- Dark tourism

Terrorism

This is the latest accepted revision, reviewed on *27 November 2016*.

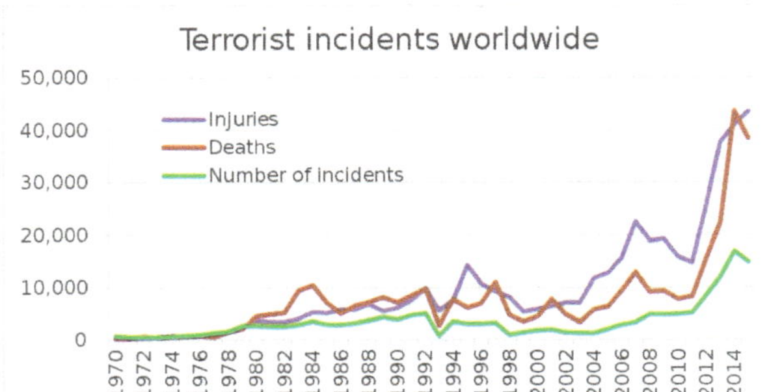

Worldwide non-state terrorist incidents 1970–2015

Terrorism is, in its broadest sense, the use of intentionally indiscriminate violence (terror) in order to achieve a political, religious, or ideological aim. It is classified as fourth-generation warfare and as a violent crime. In modern times, terrorism is considered a major threat to society and therefore illegal under anti-terrorism laws in most jurisdictions. It is also considered a war crime under the laws of war when used to target non-combatants, such as civilians, neutral military personnel, or enemy prisoners of war.

A broad array of political organizations have practiced terrorism to further their objectives. It has been practiced by both right-wing and left-wing political organizations, nationalist groups, religious groups, revolutionaries, and ruling governments. The symbolism of terrorism can exploit human fear to help achieve these goals.

According to data from the Global Terrorism Database, more than 61,000 incidents of non-state terrorism claiming over 140,000 lives have been recorded from 2000 to 2014.

Terminology

Origin of Term

Mass killings in the Vendée during the Reign of Terror in France, 1793

"Terrorism" comes from the French word *terrorisme*, and originally referred specifically to state terrorism as practiced by the French government during the 1793–1794 Reign of Terror. The French word *terrorisme* in turn derives from the Latin verb *terrere* (e, terreo) meaning "to frighten". The *terror cimbricus* was a panic and state of emergency in Rome in response to the approach of warriors of the Cimbri tribe in 105 BCE that the Jacobins cited as a precedent when imposing the Reign of Terror during the French Revolution. After the Jacobins lost power, the word "terrorist" became a term of abuse.

Although "terrorism" originally referred to acts committed by a government, currently it usually refers to the killing of innocent people for political purposes in such a way as to create a spectacle. This meaning can be traced back to Sergey Nechayev, who described himself as a "terrorist". Nechayev founded the Russian terrorist group "People's Retribution" in 1869.

The lack of consensus as to what a terrorist is can affect policies designed to deal with terrorists. Some view them as soldiers that can be held at the end of a war and are entitled to various privileges spelled out in the Geneva Conventions. Others view them as criminals that should be tried in civil courts. Still others will argue that terrorists are best treated as a category to themselves and need policies tailored to them.

In November 2004, a Secretary-General of the United Nations report described terrorism as any act "intended to cause death or serious bodily harm to civilians or non-combatants with the purpose of intimidating a population or compelling a government or an international organization to do or abstain from doing any act".

Definition

Attack at the Bologna railway station on 2 August 1980 by the neo-fascist group Nuclei Armati Rivoluzionari. With 85 deaths, it is the deadliest massacre in the history of Italy as a Republic.

The definition of terrorism has proven controversial. Various legal systems and government agencies use different definitions of terrorism in their national legislation. Moreover, the international community has been slow to formulate a universally agreed, legally binding definition of this crime. These difficulties arise from the fact that the term "terrorism" is politically and emotionally charged. In this regard, Angus Martyn, briefing the Australian Parliament, stated,

The international community has never succeeded in developing an accepted comprehensive definition of terrorism. During the 1970s and 1980s, the United Nations attempts to define the term floundered mainly due to differences of opinion between various members about the use of violence in the context of conflicts over national liberation and self-determination.

These divergences have made it impossible for the United Nations to conclude a Comprehensive Convention on International Terrorism that incorporates a single, all-encompassing, legally binding, criminal law definition of terrorism. The international community has adopted a series of sectoral conventions that define and criminalize various types of terrorist activities.

Since 1994, the United Nations General Assembly has repeatedly condemned terrorist acts using the following political description of terrorism:

Criminal acts intended or calculated to provoke a state of terror in the public, a group of persons or particular persons for political purposes are in any circumstance unjustifiable, whatever the considerations of a political, philosophical, ideological, racial, ethnic, religious or any other nature that may be invoked to justify them.

U.S. Code Title 22 Chapter 38, Section 2656f(d) defines terrorism as: "Premeditated, politically motivated violence perpetrated against noncombatant targets by subnational groups or clandestine agents, usually intended to influence an audience."

Bruce Hoffman, a scholar, has noted:

It is not only individual agencies within the same governmental apparatus that cannot agree on a single definition of terrorism. Experts and other long-established scholars in the field are equally incapable of reaching a consensus. In the first edition of his magisterial survey, 'Political Terrorism: A Research Guide,' Alex Schmid devoted more than a hundred pages to examining more than a hundred different definitions of terrorism in an effort to discover a broadly acceptable, reasonably comprehensive explication of the word. Four years and a second edition later, Schmid was no closer to the goal of his quest, conceding in the first sentence of the revised volume that the "search for an adequate definition is still on". Walter Laqueur despaired of defining terrorism in both editions of his monumental work on the subject, maintaining that it is neither possible to do so nor worthwhile to make the attempt.

Hoffman believes it is possible to identify some key characteristics of terrorism. He proposes that:

The Baghdad bus station was the scene of a triple car bombing in August 2005 that killed 43 people.

By distinguishing terrorists from other types of criminals and terrorism from other forms of crime, we come to appreciate that terrorism is :

- ineluctably political in aims and motives;
- violent – or, equally important, threatens violence;
- designed to have far-reaching psychological repercussions beyond the immediate victim or target;
- conducted either by an organization with an identifiable chain of command or conspiratorial cell structure (whose members wear no uniform or identifying insignia) or by individuals or a small collection of individuals directly influenced, motivated, or inspired by the ideological aims or example of some existent terrorist movement and/or its leaders;

and

- perpetrated by a subnational group or nonstate entity.

A definition proposed by Carsten Bockstette at the George C. Marshall European Center for Security Studies, underlines the psychological and tactical aspects of terrorism:

Terrorism is defined as political violence in an asymmetrical conflict that is designed to induce terror and psychic fear (sometimes indiscriminate) through the violent victimization and destruction of noncombatant targets (sometimes iconic symbols). Such acts are meant to send a message from an illicit clandestine organization. The purpose of terrorism is to exploit the media in order to achieve maximum attainable publicity as an amplifying force multiplier in order to influence the targeted audience(s) in order to reach short- and midterm political goals and/or desired long-term end states.

Oslo, Norway immediately after the 2011 terrorist attack in Norway perpetrated by Anders Behring Breivik

Each act of terrorism is a "performance" devised to affect many large audiences. Terrorists also attack national symbols, to show power and to attempt to shake the foundation of the country or society they are opposed to. This may negatively affect a government, while increasing the prestige of the given terrorist group and/or ideology behind a terrorist act.

Terrorist acts frequently have a political purpose. This is often where the inter-relationship between terrorism and religion occurs. When a political struggle is integrated into the framework of a religious or "cosmic" struggle, such as over the control of an ancestral homeland or holy site such as Israel and Jerusalem, failing in the political goal (nationalism) becomes equated with spiritual failure, which, for the highly committed, is worse than their own death or the deaths of innocent civilians.

Their suffering accomplishes the terrorists' goals of instilling fear, getting their message out to an audience or otherwise satisfying the demands of their often radical religious and political agendas.

Some official, governmental definitions of terrorism use the criterion of the illegitimacy or unlawfulness of the act. to distinguish between actions authorized by a government (and thus "lawful") and those of other actors, including individuals and small groups. For example, carrying out a strategic bombing on an enemy city, which is designed to affect civilian support for a cause, would not be considered terrorism if it were authorized by a government. This criterion is inherently problematic and is not universally accepted, because: it denies the existence of state terrorism; the same act may or may not be classed as terrorism depending on whether its sponsorship is traced to a "legitimate" government; "legitimacy" and "lawfulness" are subjective, depending on the perspective of one government or another; and it diverges from the historically accepted meaning and origin of the term.

According to Ali Khan, the distinction lies ultimately in a political judgment.

An associated, and arguably more easily definable, but *not equivalent* term is violent non-state actor. The semantic scope of this term includes not only "terrorists", but while excluding some individuals or groups who have previously been described as "terrorists", and also explicitly excludes state terrorism.

Barack Obama, commenting on the Boston Marathon bombings of April 2013, declared that "[a] nytime bombs are used to target innocent civilians, it is an act of terror." Various commentators have pointed out the distinction between "act of terror" and "terrorism", particularly when used by the White House. 18 U.S.C. § 2331 defines "international terrorism" and "domestic terrorism" for purposes of Chapter 113B of the Code, entitled "Terrorism":

"International terrorism" means activities with the following three characteristics:

Involve violent acts or acts dangerous to human life that violate federal or state law; Appear to be intended (i) to intimidate or coerce a civilian population; (ii) to influence the policy of a government by intimidation or coercion; or (iii) to affect the conduct of a government by mass destruction, assassination, or kidnapping; and Occur primarily outside the territorial jurisdiction of the U.S., or transcend national boundaries in terms of the means by which they are accomplished, the persons they appear intended to intimidate or coerce, or the locale in which their perpetrators operate or seek asylum.

Pejorative Use

Those labeled "terrorists" by their opponents rarely identify themselves as such, and typically use other terms or terms specific to their situation, such as separatist, freedom fighter, liberator, revolutionary, vigilante, militant, paramilitary, guerrilla, rebel, patriot, or any similar-meaning word in other languages and cultures. Jihadi, mujaheddin, and fedayeen are similar Arabic words that have entered the English lexicon. It is common for both parties in a conflict to describe each other as terrorists.

On whether particular terrorist acts, such as killing non-combatants, can be justified as the lesser evil in a particular circumstance, philosophers have expressed different views: while, according to David Rodin, utilitarian philosophers can (in theory) conceive of cases in which the evil of terrorism is outweighed by the good that could not be achieved in a less morally costly way, in practice the "harmful effects of undermining the convention of non-combatant immunity is thought to outweigh the goods that may be achieved by particular acts of terrorism". Among the non-utilitarian philosophers, Michael Walzer argued that terrorism can be morally justified in only one specific case: when "a nation or community faces the extreme threat of complete destruction and the only way it can preserve itself is by intentionally targeting non-combatants, then it is morally entitled to do so".

In his book *Inside Terrorism* Bruce Hoffman offered an explanation of why the term *terrorism* becomes distorted:

On one point, at least, everyone agrees: *terrorism* is a pejorative term. It is a word with intrinsically negative connotations that is generally applied to one's enemies and opponents, or to those with whom one disagrees and would otherwise prefer to ignore. 'What is called terrorism,' Brian Jenkins has written, 'thus seems to depend on one's point of view. Use of the term implies a moral judgment; and if one party can successfully attach the label *terrorist* to its opponent, then it has indirectly persuaded others to adopt its moral viewpoint.' Hence the decision to call someone or label some organization *terrorist* becomes almost unavoidably subjective, depending largely on whether one sympathizes with or opposes the person/group/cause concerned. If one identifies with the victim of the violence, for example, then the act is terrorism. If, however, one identifies with the perpetrator, the violent act is regarded in a more sympathetic, if not positive (or, at the worst, an ambivalent) light; and it is not terrorism.

President Reagan meeting with Afghan Mujahideen leaders in the Oval Office in 1983

The pejorative connotations of the word can be summed up in the aphorism, "One man's terrorist is another man's freedom fighter". This is exemplified when a group using irregular military methods is an ally of a state against a mutual enemy, but later falls out with the state and starts to use those methods against its former ally. During World War II, the Malayan People's Anti-Japanese Army was allied with the British, but during the Malayan Emergency, members of its successor (the Malayan Races Liberation Army), were branded "terrorists" by the British. More recently, Ronald Reagan and others in the American administration frequently called the mujaheddin "freedom fighters" during the Soviet–Afghan War yet twenty years later, when a new generation of Afghan men were fighting against what they perceive to be a regime installed by foreign powers, their attacks were labelled "terrorism" by George W. Bush. Groups accused of terrorism understandably prefer terms reflecting legitimate military or ideological action. Leading terrorism researcher Professor Martin Rudner, director of the Canadian Centre of Intelligence and Security Studies at Ottawa's Carleton University, defines "terrorist acts" as unlawful attacks for political or other ideological goals, and said:

There is the famous statement: 'One man's terrorist is another man's freedom fighter.' But that is grossly misleading. It assesses the validity of the cause when terrorism is an act. One can have a perfectly beautiful cause and yet if one commits terrorist acts, it is terrorism regardless.

Some groups, when involved in a "liberation" struggle, have been called "terrorists" by the Western governments or media. Later, these same persons, as leaders of the liberated nations, are called "statesmen" by similar organizations. Two examples of this phenomenon are the Nobel Peace Prize laureates Menachem Begin and Nelson Mandela. WikiLeaks editor Julian Assange has been called a "terrorist" by Sarah Palin and Joe Biden.

Sometimes, states that are close allies, for reasons of history, culture and politics, can disagree over whether or not members of a certain organization are terrorists. For instance, for many years, some branches of the United States government refused to label members of the Provisional Irish Republican Army (IRA) as terrorists while the IRA was using methods against one of the United States' closest allies (the United Kingdom) that the UK branded as terrorism. This was highlighted by the *Quinn v. Robinson* case.

Media outlets who wish to convey impartiality may limit their usage of "terrorist" and "terrorism" because they are loosely defined, potentially controversial in nature, and subjective terms.

History

The Irish Republican Brotherhood was one of the earliest organizations to use modern terrorist tactics. Pictured, "The Fenian Guy Fawkes" by John Tenniel (1867).

Depending on how broadly the term is defined, the roots and practice of terrorism can be traced at least to the 1st-century AD. Sicarii Zealots, though some dispute whether the group, a radical offshoot of the Zealots which was active in Judaea Province at the beginning of the 1st century AD, was in fact terrorist. According to the contemporary Jewish-Roman historian Josephus, after the Zealotry rebellion against Roman rule in Judea, when some prominent Jewish collaborators with Roman rule were killed, Judas of Galilee formed a small and more extreme offshoot of the Zealots, the Sicarii, in 6 AD. Their terror was also directed against Jewish "collaborators", including temple priests, Sadducees, Herodians, and other wealthy elites.

The term "terrorism" itself was originally used to describe the actions of the Jacobin Club during the "Reign of Terror" in the French Revolution. "Terror is nothing other than justice, prompt, severe, inflexible," said Jacobin leader Maximilien Robespierre. In 1795, Edmund Burke denounced the Jacobins for letting "thousands of those hell-hounds called Terrorists ... loose on the people" of France.

In January 1858, Italian patriot Felice Orsini threw three bombs in an attempt to assassinate French Emperor Napoleon III. Eight bystanders were killed and 142 injured. The incident played a crucial role as an inspiration for the development of the early terrorist groups.

Arguably the first organization to utilize modern terrorist techniques was the Irish Republican Brotherhood, founded in 1858 as a revolutionary Irish nationalist group that carried out attacks in England. The group initiated the Fenian dynamite campaign in 1881, one of the first modern terror campaigns. Instead of earlier forms of terrorism based on political assassination, this campaign used modern, timed explosives with the express aim of sowing fear in the very heart of metropolitan Britain, in order to achieve political gains.

Another early terrorist group was Narodnaya Volya, founded in Russia in 1878 as a revolutionary anarchist group inspired by Sergei Nechayev and "propaganda by the deed" theorist Pisacane. The group developed ideas—such as targeted killing of the 'leaders of oppression'—that were to become the hallmark of subsequent violence by small non-state groups, and they were convinced that the developing technologies of the age—such as the invention of dynamite, which they were the first anarchist group to make widespread use of—enabled them to strike directly and with discrimination. Modern terrorism had largely taken shape by the turn of the 20th century.

Infographics

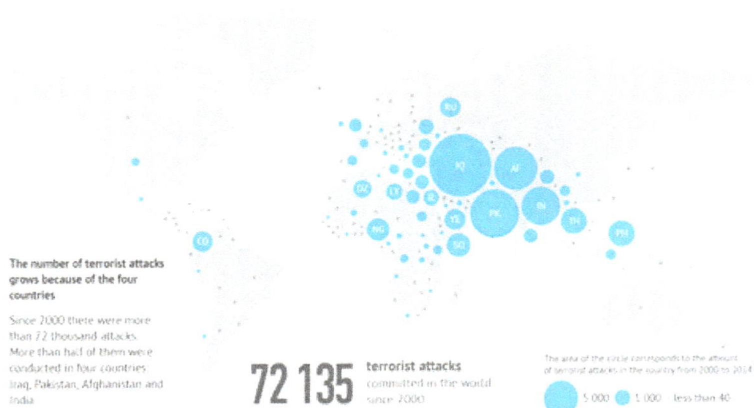

The number of terrorist attacks worldwide from 2000 to 2014

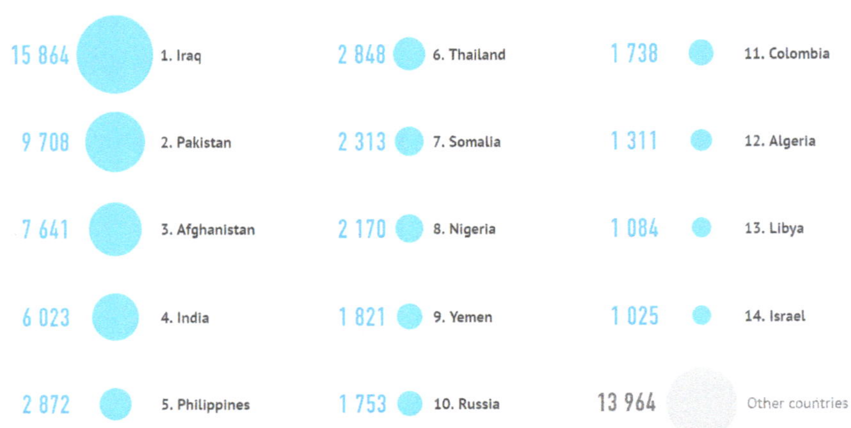

Top 10 Countries (2000–2014)

Types

Depending on the country, the political system, and the time in history, the types of terrorism are varying.

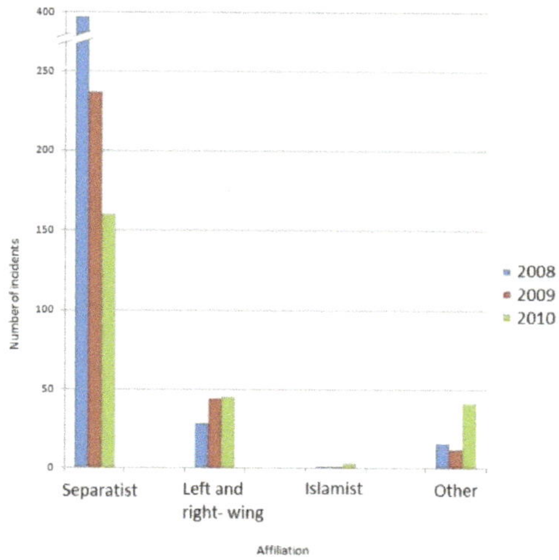

Number of failed, foiled or successful terrorist attacks by year and type within the European Union. Source: Europol. 1 person died in terrorist attacks from separatist groups in 2010.

Aftermath of the King David Hotel bombing by the Zionist militant group Irgun, July 1946

A view of damage to the U.S. Embassy in the aftermath of the 1983 Beirut bombing caused by Islamic Jihad Organization and Hezbollah

In early 1975, the Law Enforcement Assistant Administration in the United States formed the National Advisory Committee on Criminal Justice Standards and Goals. One of the five volumes that the committee wrote was titled *Disorders and Terrorism*, produced by the Task Force on Disorders and Terrorism under the direction of H. H. A. Cooper, Director of the Task Force staff.

The Task Force defines terrorism as "a tactic or technique by means of which a violent act or the threat thereof is used for the prime purpose of creating overwhelming fear for coercive purposes." It classified disorders and terrorism into six categories:

- Civil disorder – A form of collective violence interfering with the peace, security, and normal functioning of the community.
- Political terrorism – Violent criminal behaviour designed primarily to generate fear in the community, or substantial segment of it, for political purposes.
- Non-Political terrorism – Terrorism that is not aimed at political purposes but which exhibits "conscious design to create and maintain a high degree of fear for coercive purposes, but the end is individual or collective gain rather than the achievement of a political objective."
- Quasi-terrorism – The activities incidental to the commission of crimes of violence that are similar in form and method to genuine terrorism but which nevertheless lack its essential ingredient. It is not the main purpose of the quasi-terrorists to induce terror in the immediate victim as in the case of genuine terrorism, but the quasi-terrorist uses the modalities and techniques of the genuine terrorist and produces similar consequences and reaction. For example, the fleeing felon who takes hostages is a quasi-terrorist, whose methods are similar to those of the genuine terrorist but whose purposes are quite different.
- Limited political terrorism – Genuine political terrorism is characterized by a revolutionary approach; limited political terrorism refers to "acts of terrorism which are committed for ideological or political motives but which are not part of a concerted campaign to capture control of the state."
- Official or state terrorism – "referring to nations whose rule is based upon fear and oppression that reach similar to terrorism or such proportions." It may also be referred to as Structural Terrorism defined broadly as terrorist acts carried out by governments in pursuit of political objectives, often as part of their foreign policy.

Other sources have defined the typology of terrorism in different ways, for example, broadly classifying it into domestic terrorism and international terrorism, or using categories such as vigilante terrorism or insurgent terrorism. One way the typology of terrorism may be defined:

- Political terrorism
    - Sub-state terrorism
        - Social revolutionary terrorism
        - Nationalist-separatist terrorism
        - Religious extremist terrorism

- ☐ Religious fundamentalist Terrorism
- ☐ New religions terrorism
- ☐ Right-wing terrorism
- ☐ Left-wing terrorism
- o State-sponsored terrorism
- o Regime or state terrorism
- Criminal terrorism
- Pathological terrorism

Motivations of Terrorists

Attacks on 'collaborators' are used to intimidate people from cooperating with the state in order to undermine state control. This strategy was used in Ireland, in Kenya, in Algeria and in Cyprus during their independence struggles.

Attacks on high-profile symbolic targets are used to incite counter-terrorism by the state to polarize the population. This strategy was used by Al-Qaeda in its attacks on the World Trade Center and the Pentagon in the United States on September 11, 2001. These attacks are also used to draw international attention to struggles that are otherwise unreported, such as the Palestinian airplane hijackings in 1970 and the South Moluccan hostage crisis in the Netherlands in 1975.

Abrahm suggests that terrorist organizations do not select terrorism for its political effectiveness. Individual terrorists tend to be motivated more by a desire for social solidarity with other members of their organization than by political platforms or strategic objectives, which are often murky and undefined. Additionally, Michael Mousseau shows possible relationships between the type of economy within a country and ideology associated with terrorism.

Some terrorists like Timothy McVeigh were motivated by revenge against a state for its actions against its citizens.

Democracy and Domestic Terrorism

Demonstration in Madrid against ETA, January 2000. Roughly a million people met there.

The relationship between domestic terrorism and democracy is very complex. Terrorism is most common in nations with intermediate political freedom, and it is least common in the most democratic nations. However, one study suggests that Suicide attack may be an exception to this general rule. Evidence regarding this particular method of terrorism reveals that every modern suicide campaign has targeted a democracy–a state with a considerable degree of political freedom. The study suggests that concessions awarded to terrorists during the 1980s and 1990s for suicide attacks increased their frequency. There is a connection between the existence of civil liberties, democratic participation and terrorism. According to Young and Dugan, these things encourage terrorist groups to organize and generate terror.

Some examples of "terrorism" in non-democratic nations include ETA in Spain under Francisco Franco (although the group's terrorist activities increased sharply after Franco's death), the Organization of Ukrainian Nationalists in pre-war Poland, the Shining Path in Peru under Alberto Fujimori, the Kurdistan Workers Party when Turkey was ruled by military leaders and the ANC in South Africa. Democracies, such as Japan, the United Kingdom, the United States, Israel, Indonesia, India, Spain, Germany and the Philippines, have also experienced domestic terrorism.

While a democratic nation espousing civil liberties may claim a sense of higher moral ground than other regimes, an act of terrorism within such a state may cause a dilemma: whether to maintain its civil liberties and thus risk being perceived as ineffective in dealing with the problem; or alternatively to restrict its civil liberties and thus risk delegitimizing its claim of supporting civil liberties. For this reason, homegrown terrorism has started to be seen as a greater threat, as stated by former CIA Director Michael Hayden. This dilemma, some social theorists would conclude, may very well play into the initial plans of the acting terrorist(s); namely, to delegitimize the state and cause a systematic shift towards anarchy via the accumulation of negative sentiments towards the state system.

Religious Terrorism

Civilians trapped in a London Underground train after a bomb exploded further down the train at Russell Square Tube station on 7 July 2005

Islamabad Marriott Hotel bombing. Some 35,000 Pakistanis have died from terrorist attacks in recent years.

Religious terrorism is terrorism performed by groups or individuals, the motivation of which is typically rooted in faith-based tenets. Terrorist acts throughout history have been performed on religious grounds with the goal to either spread or enforce a system of belief, viewpoint or opinion. The validity and scope of religious terrorism is limited to the individual or a group view or interpretation of that belief system's teachings.

Intimate Terrorism

Intimate terrorism (IT) may also involve emotional and psychological abuse. Intimate terrorism is one element in a general pattern of control by one partner over the other. Intimate terrorism is more likely to escalate over time, not as likely to be mutual, and more likely to involve serious injury. IT batterers include two types: "Generally-violent-antisocial" and "dysphoric-borderline". The first type includes people with general psychopathic and violent tendencies. The second type are people who are emotionally dependent on the relationship. Violence by a person against their intimate partner is often done as a way for controlling their partner, even if this kind of violence is not the most frequent. Support for this typology has been found in subsequent evaluations.

Perpetrators

The perpetrators of acts of terrorism can be individuals, groups, or states. According to some definitions, clandestine or semi-clandestine state actors may also carry out terrorist acts outside the framework of a state of war. However, the most common image of terrorism is that it is carried out by small and secretive cells, highly motivated to serve a particular cause and many of the most deadly operations in recent times, such as the September 11 attacks, the London underground bombing, 2008 Mumbai attacks and the 2002 Bali bombing were planned and carried out by a close clique, composed of close friends, family members and other strong social networks. These groups benefited from the free flow of information and efficient telecommunications to succeed where others had failed.

Over the years, much research has been conducted to distill a terrorist profile to explain these individuals' actions through their psychology and socio-economic circumstances. Others, like Roderick Hindery, have sought to discern profiles in the propaganda tactics used by terrorists. Some security organizations designate these groups as *violent non-state actors*. A 2007 study by economist Alan B. Krueger found that terrorists were less likely to come from an impoverished background (28% vs. 33%) and more likely to have at least a high-school education (47% vs. 38%). Another analysis found only 16% of terrorists came from impoverished families, vs. 30% of male Palestinians, and over 60% had gone beyond high school, vs. 15% of the populace.

To avoid detection, a terrorist will look, dress, and behave normally until executing the assigned mission. Some claim that attempts to profile terrorists based on personality, physical, or sociological traits are not useful. The physical and behavioral description of the terrorist could describe almost any normal person. However, the majority of terrorist attacks are carried out by military age men, aged 16–40.

Non-state Groups

There is speculation that anthrax mailed inside letters to U.S. politicians was the work of a *lone wolf*.

Groups not part of the state apparatus of in opposition to the state are most commonly referred to as a "terrorist" in the media.

State Sponsors

A state can sponsor terrorism by funding or harboring a terrorist group. Opinions as to which acts of violence by states consist of state-sponsored terrorism vary widely. When states provide funding for groups considered by some to be terrorist, they rarely acknowledge them as such.

State Terrorism

Civilization is based on a clearly defined and widely accepted yet often unarticulated hierarchy. Violence done by those higher on the hierarchy to those lower is nearly always invisible, that is, unnoticed. When it is noticed, it is fully rationalized. Violence done by those lower on the hierarchy to those higher is unthinkable, and when it does occur is regarded with shock, horror, and the fetishization of the victims.

—Derrick Jensen

Infant crying in Shanghai's South Station after the Japanese bombing, August 28, 1937.

As with "terrorism" the concept of "state terrorism" is controversial. The Chairman of the United Nations Counter-Terrorism Committee has stated that the Committee was conscious of 12 international Conventions on the subject, and none of them referred to State terrorism, which was not an international legal concept. If States abused their power, they should be judged against international conventions dealing with war crimes, international human rights law, and international humanitarian law. Former United Nations Secretary-General Kofi Annan has said that it is "time to set aside debates on so-called 'state terrorism'. The use of force by states is already thoroughly regulated under international law". However, he also made clear that, "regardless of the differences between governments on the question of definition of terrorism, what is clear and what we can all agree on is any deliberate attack on innocent civilians [or non-combatants], regardless of one's cause, is unacceptable and fits into the definition of terrorism."

USS Arizona (BB-39) burning during the Japanese surprise attack on Pearl Harbor, December 7, 1941.

State terrorism has been used to refer to terrorist acts committed by governmental agents or forces. This involves the use of state resources employed by a state's foreign policies, such as using its military to directly perform acts of terrorism. Professor of Political Science Michael Stohl cites the examples that include the German bombing of London, the Japanese bombing of Pearl Harbor, the British firebombing of Dresden, and the U.S. atomic bombing of Hiroshima during World War II. He argues that "the use of terror tactics is common in international relations and the state has been and remains a more likely employer of terrorism within the international system than insurgents." He also cites the first strike option as an example of the "terror of coercive diplomacy" as a form of this, which holds the world hostage with the implied threat of using nuclear weapons in "crisis management" and argue that the institutionalized form of terrorism has occurred as a result of changes that took place following World War II. In this analysis, state terrorism exhibited as a form of foreign policy was shaped by the presence and use of weapons of mass destruction, and that the legitimizing of such violent behavior led to an increasingly accepted form of this state behavior.

St Paul's Cathedral after the German bombing of London, c. 1940.

Charles Stewart Parnell described William Ewart Gladstone's Irish Coercion Act as terrorism in his "no-Rent manifesto" in 1881, during the Irish Land War. The concept is also used to describe political repressions by governments against their own civilian populations with the purpose of inciting fear. For example, taking and executing civilian hostages or extrajudicial elimination campaigns are commonly considered "terror" or terrorism, for example during the Red Terror or the Great Terror. Such actions are also often described as democide or genocide, which have been argued to be equivalent to state terrorism. Empirical studies on this have found that democracies have little democide.

Connection with Tourism

The connection between terrorism and tourism has been widely studied since the Luxor massacre in Egypt. In the 1970s, the targets of terrorists were politicians and chiefs of police while now international tourists and visitors are selected as main targets of attacks. The attacks on the World Trade Center and the Pentagon on September 11, 2001, were the symbolic epicenter, which marked a new epoch in the use of civil transport against the main power of the planet. From this event onwards, the spaces of leisure that characterized the pride of West, were conceived as dangerous and frightful. Maximiliano E Korstanje argued that terrorism represents a dialectic of hate, between a group of insurgents whose interests have been placed outside the electoral system and the state which is unable to anticipate the next blow. Historically, tourism and terrorism have inextricably intertwined. As enrooted in the capitalist ethos, terrorism rests on the logic of violence and extortion, where outsiders are used to achieve the in-group's goals. Similarly, Luke Howie explains that the actions of terrorists are not aimed at effacing entire civilizations, as the media portrays, but in administering an extreme fear so that their claims will be accepted. Terrorists are usually psychologically insensitive to the suffering of others. Using extortion as a main tactic, the media lays a fertile ground which amplifies the effects of terrorism on the society. Likely, one of the main problems of terrorism seems to be the need to capture the attention of an audience. To some extent, terrorists appear to jolt the society, however, the western audience experiences a gradual process of desensitization. This result leads these groups to innovate more cruel and violent strategies.

Funding

State sponsors have constituted a major form of funding; for example, Palestine Liberation Organization, Democratic Front for the Liberation of Palestine and other groups considered to be terrorist organizations, were funded by the Soviet Union. The Stern Gang received funding from Italian Fascist officers in Beirut to undermine the British Mandate for Palestine. Pakistan has created and nurtured terrorist groups as policy for achieving tactical objectives against its neighbours, especially India.

"Revolutionary tax" is another major form of funding, and essentially a euphemism for "protection money". Revolutionary taxes are typically extorted from businesses (including farms cultivating illicit drugs (such as Papaver somniferum)) and they also "play a secondary role as one other means of intimidating the target population".

Other major sources of funding include kidnapping for ransoms, smuggling (including wildlife smuggling), fraud, and robbery. The Islamic State in Iraq and the Levant has reportedly received funding "via private donations from the Gulf states".

The Financial Action Task Force is an inter-governmental body whose mandate, since October 2001, has included combatting terrorist financing.

Tactics

The Wall Street bombing at noon on September 16, 1920 killed thirty-eight people and injured several hundred. The perpetrators were never caught.

Terrorism is a form of asymmetric warfare, and is more common when direct conventional warfare will not be effective because forces vary greatly in power.

The context in which terrorist tactics are used is often a large-scale, unresolved political conflict. The type of conflict varies widely; historical examples include:

- Secession of a territory to form a new sovereign state or become part of a different state
- Dominance of territory or resources by various ethnic groups
- Imposition of a particular form of government
- Economic deprivation of a population

- Opposition to a domestic government or occupying army
- Religious fanaticism

Terrorist attacks are often targeted to maximize fear and publicity, usually using explosives or poison. There is concern about terrorist attacks employing weapons of mass destruction. Terrorist groups usually methodically plan attacks in advance, and may train participants, plant undercover agents, and raise money from supporters or through organized crime. Communications occur through modern telecommunications, or through old-fashioned methods such as couriers.

Responses

Sign notifying shoppers of increased surveillance due to a perceived increased risk of terrorism

Responses to terrorism are broad in scope. They can include re-alignments of the political spectrum and reassessments of fundamental values.

Specific types of responses include:

- Targeted laws, criminal procedures, deportations, and enhanced police powers
- Target hardening, such as locking doors or adding traffic barriers
- Preemptive or reactive military action
- Increased intelligence and surveillance activities
- Preemptive humanitarian activities
- More permissive interrogation and detention policies

The term "counter-terrorism" has a narrower connotation, implying that it is directed at terrorist actors.

Response in the United States

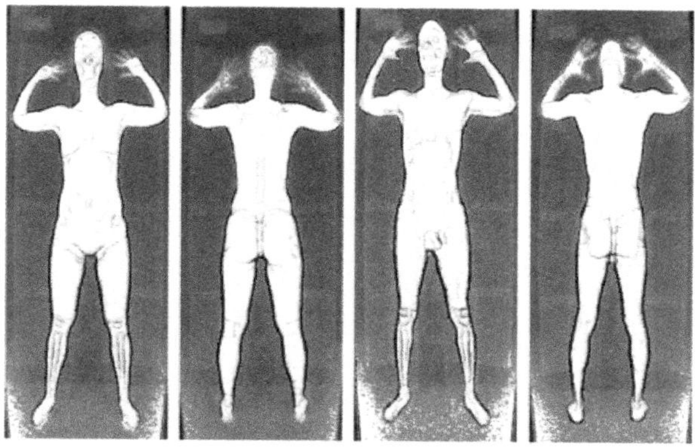

X-ray backscatter technology (AIT) machine used by the TSA to screen passengers. According to the TSA, this is what the remote TSA agent would see on their screen.

According to a report by Dana Priest and William M. Arkin in The Washington Post, "Some 1,271 government organizations and 1,931 private companies work on programs related to counterterrorism, homeland security and intelligence in about 10,000 locations across the United States."

America's thinking on how to defeat radical Islamists is split along two very different schools of thought. Republicans, typically follow what is known as the Bush Doctrine, advocate the military model of taking the fight to the enemy and seeking to democratize the Middle East. Democrats, by contrast, generally propose the law enforcement model of better cooperation with nations and more security at home. In the introduction of the *U.S. Army / Marine Corps Counterinsurgency Field Manual,* Sarah Sewall states the need for "U.S. forces to make securing the civilian, rather than destroying the enemy, their top priority. The civilian population is the center of gravity—the deciding factor in the struggle.... Civilian deaths create an extended family of enemies—new insurgent recruits or informants—–and erode support of the host nation." Sewall sums up the book's key points on how to win this battle: "Sometimes, the more you protect your force, the less secure you may be.... Sometimes, the more force is used, the less effective it is.... The more successful the counterinsurgency is, the less force can be used and the more risk must be accepted.... Sometimes, doing nothing is the best reaction." This strategy, often termed "courageous restraint," has certainly led to some success on the Middle East battlefield, yet it fails to address the central truth: the terrorists we face are mostly homegrown.

Terrorism Research

Terrorism research, also called terrorism and counter-terrorism research, is an interdisciplinary academic field which seeks to understand the causes of terrorism, how to prevent it as well as its impact in the broadest sense. Terrorism research can be carried out in both military and civilian contexts, for example by research centres such as the British Centre for the Study of Terrorism and Political Violence Centre for Violence and Traumatic Stress Studies and the International Centre for Counter-Terrorism (ICCT). There are several academic journals devoted to the field.

Mass Media

Mass media exposure may be a primary goal of those carrying out terrorism, to expose issues that would otherwise be ignored by the media. Some consider this to be manipulation and exploitation of the media.

La Terroriste, a 1910 poster depicting a female member of the Combat Organization of the Polish Socialist Party throwing a bomb at a Russian official's car.

The Internet has created a new channel for groups to spread their messages. This has created a cycle of measures and counter measures by groups in support of and in opposition to terrorist movements. The United Nations has created its own online counter-terrorism resource.

The mass media will, on occasion, censor organizations involved in terrorism (through self-restraint or regulation) to discourage further terrorism. However, this may encourage organizations to perform more extreme acts of terrorism to be shown in the mass media. Conversely James F. Pastor explains the significant relationship between terrorism and the media, and the underlying benefit each receives from the other.

There is always a point at which the terrorist ceases to manipulate the media gestalt. A point at which the violence may well escalate, but beyond which the terrorist has become symptomatic of the media gestalt itself. Terrorism as we ordinarily understand it is innately media-related.

—*Novelist William Gibson*

Former British Prime Minister Margaret Thatcher also famously spoke of the close connection between terrorism and the media, calling publicity 'the oxygen of terrorism'.

Databases

The following terrorism databases are or were made publicly available for research purposes, and track specific acts of terrorism:

- Global Terrorism Database, An open-source database on terrorist events around the world from 1970 through 2015 with more than 150,000 cases.
- MIPT Terrorism Knowledge Base
- Worldwide Incidents Tracking System
- Tocsearch (dynamic database)

The following publicly available resource indexes electronic and bibliographic resources on the subject of terrorism:

- Human Security Gateway

The following terrorism databases are maintained in secrecy by the United States Government for intelligence and counter-terrorism purposes:

- Terrorist Identities Datamart Environment
- Terrorist Screening Database

Sabotage

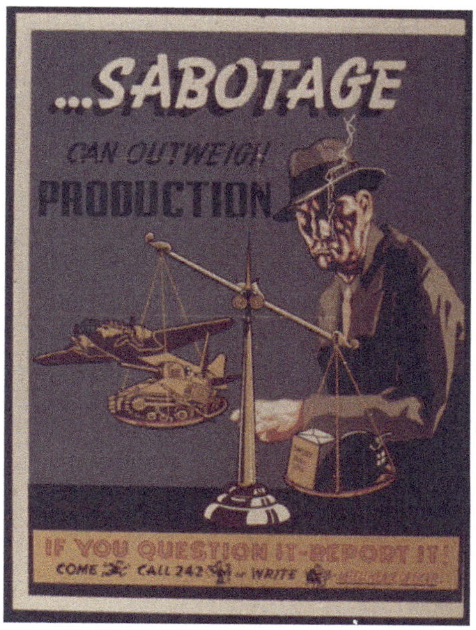

A United States poster from the World War II-era that was used to inform people about what they should do if they suspect sabotage

Sabotage is a deliberate action aimed at weakening a polity or corporation through subversion, obstruction, disruption or destruction. In a workplace setting, sabotage is the conscious withdrawal of efficiency generally directed at causing some change in workplace conditions. One who engages in sabotage is a saboteur. Saboteurs typically try to conceal their identities because of the consequences of their actions.

Any unexplained adverse condition might be sabotage. Sabotage is sometimes called tampering, meddling, tinkering, malicious pranks, malicious hacking, a practical joke or the like to avoid needing to invoke legal and organizational requirements for addressing sabotage.

Etymology

The word "sabotage" appears in the beginning of the 19th century from the French word "sabotage". It is sometimes said that some workers (from Netherlands for some, canuts from Lyon for others, luddites in England, etc.) used to throw their "sabots" (clogs) in the machines to break them, but this is not supported by the etymology.

One of its first appearances in French literature is in the *Dictionnaire du Bas-Langage ou manières de parler usitées parmi le peuple* of D'Hautel, edited in 1808.

The verb "saboter" is also found in 1873–1874 in the *Dictionnaire de la langue française* of Émile Littré. But it is at the end of the 19th century that it really began to be used with the meaning of "deliberately and maliciously destroying property" or "working slower". In 1897, Émile Pouget, a famous syndicalist and anarchist wrote "action de saboter un travail" (action of sabotaging a work) in *Le Père Peinard* and in 1911 he also wrote a book entitled *Le Sabotage*.

As Industrial Action

At the inception of the Industrial Revolution, skilled workers such as the Luddites (1811–1812) used sabotage as a means of negotiation in labor disputes.

Labor unions such as the Industrial Workers of the World (IWW) have advocated sabotage as a means of self-defense and direct action against unfair working conditions.

The IWW was shaped in part by the industrial unionism philosophy of Big Bill Haywood, and in 1910 Haywood was exposed to sabotage while touring Europe:

The experience that had the most lasting impact on Haywood was witnessing a general strike on the French railroads. Tired of waiting for parliament to act on their demands, railroad workers walked off their jobs all across the country. The French government responded by drafting the strikers into the army and then ordering them back to work. Undaunted, the workers carried their strike to the job. Suddenly, they could not seem to do anything right. Perishables sat for weeks, sidetracked and forgotten. Freight bound for Paris was misdirected to Lyon or Marseille instead. This tactic — the French called it "sabotage" — won the strikers their demands and impressed Bill Haywood.

For the IWW, sabotage came to mean any withdrawal of efficiency, including the slowdown, the strike, working to rule, or creative bungling of job assignments.

One of the most severe examples was at the construction site of the Robert-Bourassa Generating Station in 1974, in Québec, Canada, when workers used bulldozers to topple electric generators, damaged fuel tanks, and set buildings on fire. The project was delayed a year, and the direct cost of the damage estimated at $2 million CAD. The causes were not clear, but three possible factors have been cited: inter-union rivalry, poor working conditions, and the perceived arrogance of American executives of the contractor, Bechtel Corporation.

As Environmental Action

Certain groups turn to destruction of property to stop environmental destruction or to make visible arguments against forms of modern technology they consider detrimental to the environment. The U.S. Federal Bureau of Investigation (FBI) and other law enforcement agencies use the term eco-terrorist when applied to damage of property. Proponents argue that since property cannot feel terror, damage to property is more accurately described as sabotage. Opponents, by contrast, point out that property owners and operators can indeed feel terror. The image of the monkey wrench thrown into the moving parts of a machine to stop it from working was popularized by Edward Abbey in the novel *The Monkey Wrench Gang* and has been adopted by eco-activists to describe destruction of earth damaging machinery.

As War Tactic

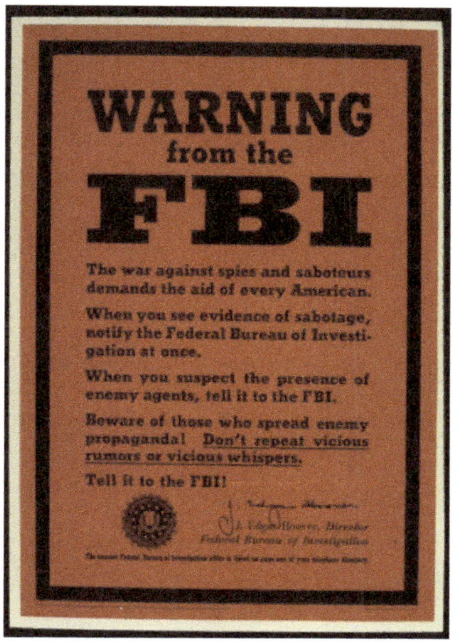

World War II poster from the United States

In war, the word is used to describe the activity of an individual or group not associated with the military of the parties at war, such as a foreign agent or an indigenous supporter, in particular when actions result in the destruction or damaging of a productive or vital facility, such as equipment, factories, dams, public services, storage plants or logistic routes. Prime examples of such sabotage are the events of Black Tom and the Kingsland Explosion. Like spies, saboteurs who conduct a military operation in civilian clothes or enemy uniforms behind enemy lines are subject to prosecution and criminal penalties instead of detention as prisoners of war. It is common for a government in power during war or supporters of the war policy to use the term loosely against opponents of the war. Similarly, German nationalists spoke of a stab in the back having cost them the loss of World War I.

A modern form of sabotage is the distribution of software intended to damage specific industrial systems. For example, the U.S. Central Intelligence Agency (CIA) is alleged to have sabotaged a Siberian pipeline during the Cold War, using information from the Farewell Dossier. A more recent case may be the Stuxnet computer worm, which was designed to subtly infect and damage specific types of industrial equipment. Based on the equipment targeted and the location of infected machines, security experts believe it was an attack on the Iranian nuclear program by the United States, Israel or, according to the latest news, even Russia.

Sabotage, done well, is inherently difficult to detect and difficult to trace to its origin. During World

War II, the U.S. Federal Bureau of Investigation (FBI) investigated 19,649 cases of sabotage and concluded the enemy had not caused any of them.

Sabotage in warfare, according to the Office of Strategic Services (OSS) manual, varies from highly technical *coup de main* acts that require detailed planning and specially trained operatives, to innumerable simple acts that ordinary citizen-saboteurs can perform. Simple sabotage is carried out in such a way as to involve a minimum danger of injury, detection, and reprisal. There are two main methods of sabotage; physical destruction and the "human element." While physical destruction as a method is self-explanatory, its targets are nuanced, reflecting objects to which the saboteur has normal and inconspicuous access in everyday life. The "human element" is based on universal opportunities to make faulty decisions, to adopt a non-cooperative attitude, and to induce others to follow suit.

There are many examples of physical sabotage in wartime. However, one of the most effective uses of sabotage is against organizations. The OSS manual provides numerous techniques under the title "General Interference with Organizations and Production":

- When possible, refer all matters to committees for "further study and consideration." Attempt to make the committees as large as possible—never fewer than five
- Bring up irrelevant issues as frequently as possible.
- Haggle over precise wordings of communications, minutes, resolutions.
- In making work assignments, always sign out unimportant jobs first, assign important jobs to inefficient workers with poor machines.
- Insist on perfect work in relatively unimportant products; send back for refinishing those with the least flaw. Approve other defective parts whose flaws are not visible to the naked eye.
- To lower morale, and with it, production, be pleasant to inefficient workers; give them undeserved promotions. Discriminate against efficient workers; complain unjustly about their work.
- Hold meetings when there is more critical work to be done.
- Multiply procedures and clearances involved in issuing instructions, pay checks, and so on.
- Spread disturbing rumors that sound like inside information.

From the section entitled, "General Devices for Lowering Morale and Creating Confusion" comes the following quintessential simple sabotage advice: "Act stupid."

Value of Simple Sabotage in Wartime

The United States Office of Strategic Services, later renamed the CIA, noted specific value in committing simple sabotage against the enemy during wartime: "... slashing tires, draining fuel tanks, starting fires, starting arguments, acting stupidly, short-circuiting electric systems, abrading machine parts will waste materials, manpower, and time." To underline the importance of simple sabotage on a widespread scale, they wrote, "Widespread practice of simple sabotage will harass and demoralize enemy administrators and police." The OSS was also focused on the battle for hearts and minds during wartime; "the very practice of simple sabotage by natives in enemy or occupied territory may make these individuals identify themselves actively with the United Nations War effort, and encourage them to assist openly in periods of Allied invasion and occupation."

In World War I

On 30 July 1916, the Black Tom explosion occurred when German agents set fire to a complex of warehouses and ships in Jersey City, New Jersey that held munitions, fuel, and explosives bound to aid the Allies in their fight.

On 11 January 1917, Fiodore Wozniak, using a rag saturated with phosphorus or an incendiary pencil supplied by German sabotage agents, set fire to his workbench at an ammunition assembly plant near Lyndhurst, New Jersey, causing a four-hour fire that destroyed half a million 3-inch explosive shells and destroyed the plant for an estimated at $17 million in damages. Wozniak's involvement was not discovered until 1927.

On 12 February 1917, Bedouins allied with the British destroyed a Turkish railroad near the port of Wajh, derailing a Turkish locomotive. The Bedouins traveled by camel and used explosives to demolish a portion of track.

Post World War I

Japanese experts inspect the scene of the "railway sabotage" on the South Manchurian Railway in 1931. The "railroad sabotage" was one of the events that led to the Mukden Incident and the Japanese occupation of Manchuria.

In Ireland, the Irish Republican Army (IRA) used sabotage against the British following the Easter 1916 uprising. The IRA compromised communication lines and lines of transportation and fuel supplies. The IRA also employed passive sabotage, refusing dock and train workers to work on ships and rail cars used by the government. In 1920, agents of the IRA committed arson against at least fifteen British warehouses in Liverpool. The following year, the IRA set fire to numerous British targets again, including the Dublin Customs House, this time sabotaging most of Liverpool's firetrucks in the firehouses before lighting the matches.

In World War II

Sabotage training for the Allies consisted of teaching would-be saboteurs key components of working machinery to destroy. "Saboteurs learned hundreds of small tricks to cause the Germans big trouble. The cables in a telephone junction box ... could be jumbled to make the wrong connections when numbers were dialed. A few ounces of plastique, properly placed, could bring down a bridge, cave in a mine shaft, or collapse the roof of a railroad tunnel."

The Polish Home Army Armia Krajowa, who ended up commanding majority of the resistance organizations in Poland (even the National Forces, except Military Organization Lizard Union; Home Army also included the Polish Socialist Party – Freedom, Equality, Independence) and coordinating and aiding the Jewish Military Union as well as more reluctantly helping Jewish Combat Organization, was responsible for the greatest number of acts of sabotage in German—occupied Europe. Home Army's sabotage operations Operation Garland and Operation Ribbon are just a couple examples. In all, the Home Army damaged 6,930 locomotives, set 443 rail transports on fire, damaged over 19,000 rail cars "wagony," and blew up 38 rail bridges, just to mention the attacks against the rail roads. The Home Army was also responsible for 4,710 built-in flaws in parts for aircraft engines and 92,000 built-in flaws in artillery projectiles, among other examples of significant sabotage. In addition, over 25,000 acts of more minor sabotage were committed. As to the Lizard Union, it continued to fight against both the Germans and the Soviets; however, it did aid the Western Allies by collecting constant and detailed information on the German rail, wheeled, and horse transports. As to the Stalin's proxies, their actions lead to a great number of the Polish and Jewish hostages, mostly civilians, murdered in reprisal by the Germans. The Gwardia Ludowa destroyed around 200 German trains during the war, and indiscriminately threw hand grenades into the places frequented by Germans. Moreover, contrary to popular belief, it was the Soviet proxies that murdered greatest number of Jews during World War Two and sometimes even wrote about it in their diaries. The motivation was mostly greed. One thing must be realized; poor people in uncertain times will not try to accumulate money, but rather will save all their wealth in the form of gold, silver, and jewels. This makes poor people great targets for theft.

The French Resistance ran an extremely effective sabotage campaign against the Germans during World War II. Receiving their sabotage orders through messages over the BBC radio or by aircraft, the French used both passive and active forms of sabotage. Passive forms included losing German shipments and allowing poor quality material to pass factory inspections. Many active sabotage attempts were against critical rail lines of transportation. German records count 1,429 instances of sabotage from French Resistance forces between January 1942 and February 1943. From January through March 1944, sabotage accounted for three times the number of locomotives damaged by Allied air power.

During World War II, the Allies committed sabotage against the Peugeot truck factory. After repeated failures in Allied bombing attempts to hit the factory, a team of French Resistance fighters and Special Operations Executive (SOE) agents distracted the German guards with a game of soccer while part of their team entered the plant and destroyed machinery.

In December 1944, the Germans ran a false flag sabotage infiltration, Operation Greif, which was commanded by Waffen-SS commando Otto Skorzeny during the Battle of the Bulge. German commandos, wearing US Army uniforms, carrying US Army weapons, and using US Army vehicles, penetrated US lines to spread panic and confusion among US troops and to blow up bridges, ammunition dumps, and fuel stores and to disrupt the lines of communication. Many of the commandos were captured by the Americans. Because they were wearing US uniforms, a number of the Germans were executed as spies, either summarily or after military commissions.

After World War II

Palestine Railway's K class 2-8-4T steam locomotive and freight train on the Jaffa and Jerusalem line after being sabotaged by Jewish paramilitary forces in 1946.

From 1948 to 1960, the Malayan Communists committed numerous effective acts of sabotage against the Malaysian Government, first targeting railway bridges, then hitting larger targets such as military camps. Most of their efforts were centered around crippling Malaysia's economy and involved sabotage against trains, rubber trees, water pipes, and electric lines. The Communist's sabotage efforts were so successful that they caused backlash amongst the Malaysian population, who gradually withdrew support for the Communist movement as their livelihoods became threatened.

In Mandatory Palestine from 1945 to 1948, Jewish groups opposed British control. Though that control was to end according to the Balfour Declaration in 1948, the groups used sabotage as an opposition tactic. The Haganah focused their efforts on camps used by the British to hold refugees and radar installations that could be used to detect illegal immigrant ships. The Stern Gang and the Irgun used terrorism and sabotage against the British government and against lines of communications. In November 1946, the Irgun and Stern Gang attacked a railroad twenty-one times in a three-week period, eventually causing shell-shocked Arab railway workers to strike. The 6th Airborne Division was called in to provide security as a means of ending the strike.

In Vietnam

The Viet Cong used swimmer saboteurs often and effectively during the Vietnam War. Between 1969 and 1970, swimmer saboteurs sunk, destroyed, or damaged 77 assets of the U.S. and its allies. Viet Cong swimmers were poorly equipped but well-trained and resourceful. The swimmers provided a low-cost/low-risk option with high payoff; possible loss to the country for failure compared to the possible gains from a successful mission led to the obvious conclusion the swimmer saboteurs were a good idea.

During the Cold War

On 1 January 1984, the Cuscatlan bridge over Lempa river in El Salvador, critical to flow of commercial and military traffic, was destroyed by guerrilla forces using explosives after using mortar fire to "scatter" the bridge's guards, causing an estimated $3.7 million in required repairs, and considerably impacting on El Salvadoran business and security.

In 1982 in Honduras, a group of nine Salvadorans and Nicaraguans destroyed a main electrical power station, leaving the capital city Tegucigalpa without power for three days.

As Crime

Some criminals have engaged in acts of sabotage for reasons of extortion. For example, Klaus-Peter Sabotta sabotaged German railway lines in the late 1990s in an attempt to extort DM10 million from the German railway operator Deutsche Bahn. He is now serving a sentence of life imprisonment. In 1989, ex-Scotland Yard detective Rodney Whitchelo was sentenced to 17 years in prison for spiking Heinz baby food products in supermarkets, in an extortion attempt on the food manufacturer.

As Political Action

The term political sabotage is sometimes used to define the acts of one political camp to disrupt, harass or damage the reputation of a political opponent, usually during an electoral campaign, such as during Watergate. Smear campaigns are a commonly used tactic. The term could also describe the actions and expenditures of private entities, corporations and organizations against democratically approved or enacted laws, policies and programs.

From 1992 to late 2007 an radical environmental activist movement known as ELF or Earth Liberation Front engaged in a near constant campaign of decentralized sabotage of any construction projects near wild lands and extractive industries such as logging and even the burning down of a ski resort of Vail Colorado. ELF used sabotage tactics often in loose coordination with other environmental activist movements to physically delay or destroy threats to wild lands as the political will developed to protect the targeted wild areas that ELF engaged.

In a Coup D'etat

Sabotage is a crucial tool of the successful coup d'etat, which requires control of communications before, during, and after the coup is staged. Simple sabotage against physical communications platforms using semi-skilled technicians, or even those trained only for this task, could effectively silence the target government of the coup, leaving the information battle space open to the dominance of the coup's leaders. To underscore the effectiveness of sabotage, "A single cooperative technician will be able temporarily to put out of action a radio station which would otherwise require a full-scale assault."

Railroads, where strategically important to the regime the coup is against, are prime targets for sabotage- if a section of the track is damaged entire portions of the transportation network can be stopped until it is fixed.

Derivative Usages

Sabotage Radio

A sabotage radio was a small two-way radio designed for use by resistance movements in World War II, and after the war often used by expeditions and similar parties.

Cybotage

Arquilla and Rondfeldt, in their work entitled *Networks and Netwars*, differentiate their definition of "netwar" from a list of "trendy synonyms," including "cybotage," a portmanteau from the words "sabotage" and "cyber." They dub the practitioners of cybotage "cyboteurs" and note while all cybotage is not netwar, some netwar is cybotage.

Counter-sabotage

Counter-sabotage, defined by *Webster's Dictionary*, is "counterintelligence designed to detect and counteract sabotage." The United States Department of Defense definition, found in the *Dictionary of Military and Associated Terms*, is "action designed to detect and counteract sabotage.

In World War II

During World War II, British subject Eddie Chapman, trained by the Germans in sabotage, became a double agent for the British. The German Abwehr entrusted Chapman to destroy the British de Havilland Company's main plant which manufactured the outstanding Mosquito light bomber, but required photographic proof from their agent to verify the mission's completion. A special unit of the Royal Engineers known as the Magic Gang covered the de Havilland plant with canvas panels and scattered papier-mâché furniture and chunks of masonry around three broken and burnt giant generators. Photos of the plant taken from the air reflected devastation for the factory and a successful sabotage mission, and Chapman, as a British sabotage double-agent, fooled the Germans for the duration of the war.

Borrowed into Japanese

In Japanese, the verb saboru means to skip school or loaf on the job.

Ecotage

Ecotage (a portmanteau of the "eco-" prefix and "sabotage") is direct action by extreme environmentalist groups (such as Earth First!) in the Western world. Ecotage is often seen as indistinguishable from acts of civil disobedience or even eco-terrorism.

Cases

All damage figures below are in United States dollars. Some well-known acts of ecotage have included:

- 1998 – Arson of buildings at Vail Mountain in the United States by the Earth Liberation Front (ELF).
- March 11, 1999 – Genetically engineered potatoes uprooted at Crop and Food research centre in New Zealand.
- December 25, 1999 – In Monmouth, Oregon, fire destroys the main office of the Boise Cascade logging company costing over $1 million ($1.4 million in 2015 dollars). ELF claim responsibility.
- 2001 – Members of the ELF were prosecuted for setting off a firebomb that caused $7 million in damages ($10 million in 2015 dollars) at the University of Washington's Center for Urban Horticulture
- 2003 – On August 1, a 206-unit condominium being built in San Diego, California was burnt down causing damage in excess of $20 million ($28 million in 2015 dollars). A 12-foot banner at the scene read "If you build it, we will burn it," signed, "The E.L.F.s are mad."
- 2003 – On August 22, arsonists associated with the Earth Liberation Front attacked several car dealerships in east suburban Los Angeles, burning down a warehouse and vandalizing over 100 vehicles, most of them SUVs or Hummers (chosen for their notoriously poor fuel efficiency) and causing over $1 million in damage ($1.3 million in 2015 dollars).

"Stop Urban Sprawl" is spray-painted on a multimillion-dollar house to protest the development. Mansions in the United States are a frequent target by the ELF.

In Fiction

Ecotage was popularized in 1975 by Edward Abbey's book *The Monkey Wrench Gang*. It has also been treated in novels by T. Coraghessan Boyle (*A Friend of the Earth*), Carl Hiaasen (*Tourist Season, Sick Puppy, Hoot*), Neal Stephenson (*Zodiac: The Eco-Thriller*), and Richard Melo (*Jokerman 8*), and movies such as *Choke Canyon* (1986) and *On Deadly Ground* (1994).

Ecotage! is also the title of a 1972 humor book by Sam Love, which is the likely origin of the word.

Ecotage is mentioned in Mars trilogy of science fiction novels by Kim Stanley Robinson as a means of protest shown by the Red political party. Typically the "Reds" would destroy terraforming ventures in an effort to slow the terraforming of Mars.

The Concrete mini series *Think Like a Mountain* is centred about ecotage aimed to protect first growth forests in the Pacific Northwest.

Genocide

Genocide is intentional action to destroy a people (usually defined as an ethnic, national, racial, or religious group) in whole or in part. The United Nations Genocide Convention defines genocide as "acts committed with intent to destroy, in whole or in part, a national, ethnic, racial or religious group".

The term *genocide* was coined in response to the Jewish Holocaust. It has subsequently been applied to many other mass killings, including the Greek genocide, the Armenian genocide, the Assyrian genocide, the Ukrainian Holodomor, the 1971 Bangladesh genocide, the Cambodian genocide; and, more recently, to the Guatemalan genocide, the Kurdish genocide, the Bosnian genocide, and the Rwandan genocide.

Etymology

Genocide has become an official term used in international relations. Before 1944, various terms, including "massacre" and "crimes against humanity" were used to describe intentional, systematic killings (and in 1941, Winston Churchill described the mass killing of Russian POWs and civilians by the German army as "a crime without a name.") In 1944, a Polish-Jewish lawyer named Raphael Lemkin created the term *genocide* to describe policies of systematic murder, in particular those being carried out by the Nazis, and the word was quickly adopted by many in the international community. The word *genocide* is the combination of the Greek prefix *geno-* (meaning "tribe" or "race") and *caedere* (the Latin word for "to kill"), and is defined as a specific set of violent crimes that are committed against a certain group with the attempt to remove the entire group from existence or to destroy them.

The word *genocide* was later included as a descriptive term to the process of indictment, but not yet as a formal legal term According to Lemkin, genocide was defined as "a coordinated strategy to destroy a group of people, a process that could be accomplished through total annihilation as well as strategies that eliminate key elements of the group's basic existence, including language, culture, and economic infrastructure." He created a concept of mobilizing much of the international relations and community, to working together and preventing the occurrence of such events happening within history and the international society. Australian anthropologist Peg LeVine coined the term "ritualcide" to describe the destruction of a group's cultural identity without necessarily destroying its members.

The study of genocide has mainly been focused towards the legal aspect of the term. Formally recognizing the act of genocide as a crime involves the undergoing prosecution that begins with not only seeing genocide as outrageous past any moral standpoint, but also may be a legal liability within international relations. When genocide is looked at in a general aspect it is viewed as the deliberate killing of a certain group. Yet is commonly seen to escape the process of trial and prosecution due to the fact that genocide is more often than not committed by the officials in power of a state or area. In 1648 before the term genocide had been coined, the Peace of Westphalia was established to protect ethnic, national, racial and in some instances religious groups. During the 19th century humanitarian intervention was needed due to the fact of conflict and justification of some of the actions executed by the military.

Raphael Lemkin, in his work *Axis Rule in Occupied Europe* (1944), or possibly in 1943, coined the term "genocide" by combining Greek *genos* and Latin *caedere* ('to kill').

Lemkin defined genocide as follows:

Generally speaking, genocide does not necessarily mean the immediate destruction of a nation, except when accomplished by mass killings of all members of a nation. It is intended rather to signify a coordinated plan of different actions aiming at the destruction of essential foundations of the life of national groups, with the aim of annihilating the groups themselves. The objectives of such a plan would be the disintegration of the political and social institutions, of culture, language, national feelings, religion, and the economic existence of national groups, and the destruction of the personal security, liberty, health, dignity, and even the lives of the individuals belonging to such groups.

The preamble to the Genocide Convention (CPPCG) notes that instances of genocide have taken place throughout history, but it was not until Lemkin coined the term and the prosecution of perpetrators of the Holocaust at the Nuremberg trials that the United Nations defined the crime of genocide under international law in the Genocide Convention.

During a video interview with Raphael Lemkin for the CBS, news commentator Quincy Howe asked him about how he came to be interested in the crime of genocide. He replied: "I became interested in genocide because it happened so many times. It happened to the Armenians, then after the Armenians, Hitler took action".

Lemkin was also a close relative of genocide victims, losing 49 relatives in the Holocaust. However, his work on defining genocide as a crime dates to 1933, and it was prompted by the Simele massacre in Iraq.

The Greek author Nikos Sarantakos claims that the first official use of the term "genocide" was made not in relation to the Holocaust, but by an international committee in 1948 referring to the kidnapping of children by the communists during the Greek Civil War.

As a Crime

International Law

After the Holocaust, which had been perpetrated by the Nazi Germany and its allies prior to and during World War II, Lemkin successfully campaigned for the universal acceptance of international laws defining and forbidding genocides. In 1946, the first session of the United Nations General Assembly adopted a resolution that "affirmed" that genocide was a crime under international law, but did not provide a legal definition of the crime. In 1948, the UN General Assembly adopted the *Convention on the Prevention and Punishment of the Crime of Genocide* (CPPCG) which defined the crime of genocide for the first time.

Members of the *Sonderkommando* burn corpses of Jews in pits at Auschwitz II-Birkenau, an extermination camp

The *CPPCG* was adopted by the UN General Assembly on 9 December 1948 and came into effect on 12 January 1951 (Resolution 260 (III)). It contains an internationally recognized definition of genocide which has been incorporated into the national criminal legislation of many countries, and was also adopted by the Rome Statute of the International Criminal Court, which established the International Criminal Court (ICC)..

...any of the following acts committed with intent to destroy, in whole or in part, a national, ethnical, racial or religious group, as such:

(a) Killing members of the group;

(b) Causing serious bodily harm, or harm to mental health, to members of the group;

(c) Deliberately inflicting on the group conditions of life calculated to bring about its physical destruction in whole or in part;

(d) Imposing measures intended to prevent births within the group;

(e) Forcibly transferring children of the group to another group.

The first draft of the Convention included political killings, but these provisions were removed in a political and diplomatic compromise following objections from some countries, including the USSR, a permanent security council member. The USSR argued that the Convention's definition should follow the etymology of the term, and may have feared greater international scrutiny of its own Great Purge. Other nations feared that including political groups in the definition would invite international intervention in domestic politics. However leading genocide scholar William Schabas states: “Rigorous examination of the travaux fails to confirm a popular impression in the literature that the opposition to inclusion of political genocide was some Soviet machination. The Soviet views were also shared by a number of other States for whom it is difficult to establish any geographic or social common denominator: Lebanon, Sweden, Brazil, Peru, Venezuela, the Philippines, the Dominican Republic, Iran, Egypt, Belgium, and Uruguay. The exclusion of political groups was in fact originally promoted by a non-governmental organization, the World Jewish Congress, and it corresponded to Raphael Lemkin’s vision of the nature of the crime of genocide.”

The convention’s purpose and scope was later described by the United Nations Security Council as follows:

The Convention was manifestly adopted for humanitarian and civilizing purposes. Its objectives are to safeguard the very existence of certain human groups and to affirm and emphasize the most elementary principles of humanity and morality. In view of the rights involved, the legal obligations to refrain from genocide are recognized as *erga omnes*.

When the Convention was drafted, it was already envisaged that it would apply not only to then existing forms of genocide, but also “to any method that might be evolved in the future with a view to destroying the physical existence of a group”. As emphasized in the preamble to the Convention, genocide has marred all periods of history, and it is this very tragic recognition that gives the concept its historical evolutionary nature.

The Convention must be interpreted in good faith, in accordance with the ordinary meaning of its terms, in their context, and in the light of its object and purpose. Moreover, the text of the Convention should be interpreted in such a way that a reason and a meaning can be attributed to every word. No word or provision may be disregarded or treated as superfluous, unless this is absolutely necessary to give effect to the terms read as a whole.

Genocide is a crime under international law regardless of “whether committed in time of peace or in time of war” (art. I). Thus, irrespective of the context in which it occurs (for example, peace time, internal strife, international armed conflict or whatever the general overall situation) genocide is a punishable international crime.

— UN Commission of Experts that examined violations of international humanitarian law committed in the territory of the former Yugoslavia.

Specific Provisions

“Intent to Destroy”

In 2007 the European Court of Human Rights (ECHR), noted in its judgement on *Jorgic v. Germany* case that in 1992 the majority of legal scholars took the narrow view that “intent to destroy” in the CPPCG meant the intended physical-biological destruction of the protected group and that this was still the majority opinion. But the ECHR also noted that a minority took a broader view and did not consider biological-physical destruction was necessary as the intent to destroy a national, racial, religious or ethnic group was enough to qualify as genocide.

In the same judgement the ECHR reviewed the judgements of several international and municipal courts judgements. It noted that International Criminal Tribunal for the Former Yugoslavia and the International Court of Justice had agreed with the narrow interpretation, that biological-physical destruction was necessary for an act to qualify as genocide. The ECHR also noted that at the time of its judgement, apart from courts in Germany which had taken a broad view, that there had been few cases of genocide under other Convention States municipal laws and that "There are no reported cases in which the courts of these States have defined the type of group destruction the perpetrator must have intended in order to be found guilty of genocide".

"In Part"

Armenian Genocide victims

The phrase "in whole or in part" has been subject to much discussion by scholars of international humanitarian law. The International Criminal Tribunal for the Former Yugoslavia found in *Prosecutor v. Radislav Krstic – Trial Chamber I – Judgment – IT-98-33 (2001) ICTY8 (2 August 2001)* that Genocide had been committed. In *Prosecutor v. Radislav Krstic – Appeals Chamber – Judgment – IT-98-33 (2004) ICTY 7 (19 April 2004)* paragraphs 8, 9, 10, and 11 addressed the issue of *in part* and found that "the part must be a substantial part of that group. The aim of the Genocide Convention is to prevent the intentional destruction of entire human groups, and the part targeted must be significant enough to have an impact on the group as a whole." The Appeals Chamber goes into details of other cases and the opinions of respected commentators on the Genocide Convention to explain how they came to this conclusion.

The judges continue in paragraph 12, "The determination of when the targeted part is substantial enough to meet this requirement may involve a number of considerations. The numeric size of the targeted part of the group is the necessary and important starting point, though not in all cases the ending point of the inquiry. The number of individuals targeted should be evaluated not only in absolute terms, but also in relation to the overall size of the entire group. In addition to the numeric size of the targeted portion, its prominence within the group can be a useful consideration. If a specific part of the group is emblematic of the overall group, or is essential to its survival, that may support a finding that the part qualifies as substantial within the meaning of Article 4 [of the Tribunal's Statute]."

In paragraph 13 the judges raise the issue of the perpetrators' access to the victims: "The historical examples of genocide also suggest that the area of the perpetrators' activity and control, as well as the possible extent of their reach, should be considered. ... The intent to destroy formed by a perpetrator of genocide will always be limited by the opportunity presented to him. While this factor alone will not indicate whether the targeted group is substantial, it can—in combination with other factors—inform the analysis."

CPPCG Coming into Force

The Convention came into force as international law on 12 January 1951 after the minimum 20 countries became parties. At that time however, only two of the five permanent members of the UN Security Council were parties to the treaty: France and the Republic of China. The Soviet Union ratified in 1954, the United Kingdom in 1970, the People's Republic of China in 1983 (having replaced the Taiwan-based Republic of China on the UNSC in 1971), and the United States in 1988. This long delay in support for the Convention by the world's most powerful nations caused the Convention to languish for over four decades. Only in the 1990s did the international law on the crime of genocide begin to be enforced.

UN Security Council on Genocide

UN Security Council Resolution 1674, adopted by the United Nations Security Council on 28 April 2006, "reaffirms the provisions of paragraphs 138 and 139 of the 2005 World Summit Outcome Document regarding the responsibility to protect populations from genocide, war crimes, ethnic cleansing and crimes against humanity". The resolution committed the Council to action to protect civilians in armed conflict.

In 2008 the UN Security Council adopted resolution 1820, which noted that "rape and other forms of sexual violence can constitute war crimes, crimes against humanity or a constitutive act with respect to genocide".

Municipal Law

Since the Convention came into effect in January 1951 about 80 United Nations member states have passed legislation that incorporates the provisions of CPPCG into their municipal law.

Criticisms of the CPPCG and other Definitions of Genocide

William Schabas has suggested that a permanent body as recommended by the Whitaker Report to monitor the implementation of the Genocide Convention, and require States to issue reports on their compliance with the convention (such as were incorporated into the United Nations Optional Protocol to the Convention against Torture), would make the convention more effective.

Writing in 1998 Kurt Jonassohn and Karin Björnson stated that the CPPCG was a legal instrument resulting from a diplomatic compromise. As such the wording of the treaty is not intended to be a definition suitable as a research tool, and although it is used for this purpose, as it has an international legal credibility that others lack, other definitions have also been postulated. Jonassohn and Björnson go on to say that none of these alternative definitions have gained widespread support for various reasons.

Jonassohn and Björnson postulate that the major reason why no single generally accepted genocide definition has emerged is because academics have adjusted their focus to emphasise different periods and have found it expedient to use slightly different definitions to help them interpret events. For example, Frank Chalk and Kurt Jonassohn studied the whole of human history, while Leo Kuper and R. J. Rummel in their more recent works concentrated on the 20th century, and Helen Fein, Barbara Harff and Ted Gurr have looked at post World War II events. Jonassohn and Björnson are critical of some of these studies, arguing that they are too expansive, and conclude that the academic discipline of genocide studies is too young to have a canon of work on which to build an academic paradigm.

The exclusion of social and political groups as targets of genocide in the CPPCG legal definition has been criticized by some historians and sociologists, for example M. Hassan Kakar in his book *The Soviet Invasion and the Afghan Response, 1979–1982* argues that the international definition of genocide is too restricted, and that it should include political groups or any group so defined by the perpetrator and quotes Chalk and Jonassohn: "Genocide is a form of one-sided mass killing in which a state or other authority intends to destroy a group, as that group and membership in it are defined by the perpetrator." While there are various definitions of the term, Adam Jones states that the majority of genocide scholars consider that "intent to destroy" is a requirement for any act to be labelled genocide, and that there is growing agreement on the inclusion of the physical destruction criterion.

Barbara Harff and Ted Gurr defined genocide as "the promotion and execution of policies by a state or its agents which result in the deaths of a substantial portion of a group ...[when] the victimized groups are defined primarily in terms of their communal characteristics, i.e., ethnicity, religion or nationality." Harff and Gurr also differentiate between genocides and politicides by the characteristics by which members of a group are identified by the state. In genocides, the victimized groups are defined primarily in terms of their communal characteristics, i.e., ethnicity, religion or nationality. In politicides the victim groups are defined primarily in terms of their hierarchical position or political opposition to the regime and dominant groups. Daniel D. Polsby and Don B. Kates, Jr. state that "... we follow Harff's distinction between genocides and 'pogroms,' which she describes as 'short-lived outbursts by mobs, which, although often condoned by authorities, rarely persist.' If the violence persists for long enough, however, Harff argues, the distinction between condonation and complicity collapses."

According to R. J. Rummel, genocide has 3 different meanings. The ordinary meaning is murder by government of people due to their national, ethnic, racial, or religious group membership. The legal meaning of genocide refers to the international treaty, the *Convention on the Prevention and Punishment of the Crime of Genocide*. This also includes non-killings that in the end eliminate the group, such as preventing births or forcibly transferring children out of the group to another group. A generalized meaning of genocide is similar to the ordinary meaning but also includes government killings of political opponents or otherwise intentional murder. It is to avoid confusion regarding what meaning is intended that Rummel created the term democide for the third meaning.

Highlighting the potential for state and non-state actors to commit genocide in the 21st century, for example, in failed states or as non-state actors acquire weapons of mass destruction, Adrian Gallagher defined genocide as 'When a source of collective power (usually a state) intentionally uses its power base to implement a process of destruction in order to destroy a group (as defined by the perpetrator), in whole or in substantial part, dependent upon relative group size'. The definition upholds the centrality of intent, the multidimensional understanding of destroy, broadens the definition of group identity beyond that of the 1948 definition yet argues that a substantial part of a group has to be destroyed before it can be classified as genocide (dependent on relative group size).

A major criticism of the international community's response to the Rwandan Genocide was that it was reactive, not proactive. The international community has developed a mechanism for prosecuting the perpetrators of genocide but has not developed the will or the mechanisms for intervening in a genocide as it happens.

International Prosecution of Genocide

By ad hoc Tribunals

Nuon Chea, the Khmer Rouge's chief ideologist, before the Cambodian Genocide Tribunal on 5 December 2011.

All signatories to the CPPCG are required to prevent and punish acts of genocide, both in peace and wartime, though some barriers make this enforcement difficult. In particular, some of the signatories—namely, Bahrain, Bangladesh, India, Malaysia, the Philippines, Singapore, the United States, Vietnam, Yemen, and former Yugoslavia—signed with the proviso that no claim of genocide could be brought against them at the International Court of Justice without their consent. Despite official protests from other signatories (notably Cyprus and Norway) on the ethics and legal standing of these reservations, the immunity from prosecution they grant has been invoked from time to time, as when the United States refused to allow a charge of genocide brought against it by former Yugoslavia following the 1999 Kosovo War.

It is commonly accepted that, at least since World War II, genocide has been illegal under customary international law as a peremptory norm, as well as under conventional international law. Acts of genocide are generally difficult to establish for prosecution, because a chain of accountability must be established. International criminal courts and tribunals function primarily because the states involved are incapable or unwilling to prosecute crimes of this magnitude themselves.

Nuremberg Tribunal (1945–1946)

Because the universal acceptance of international laws which in 1948 defined and forbade genocide with the promulgation of the *Convention on the Prevention and Punishment of the Crime of Genocide* (CPPCG), those criminals who were prosecuted after the war in international courts for taking part in the Holocaust were found guilty of crimes against humanity and other more specific crimes like murder. Nevertheless, the Holocaust is universally recognized to have been a genocide and the term, that had been coined the year before by Raphael Lemkin, appeared in the indictment of the 24 Nazi leaders, Count 3, which stated that all the defendants had "conducted deliberate and systematic genocide—namely, the extermination of racial and national groups..."

International Criminal Tribunal for the Former Yugoslavia (1993 to present)

The cemetery at the Srebrenica-Potočari Memorial and Cemetery to Genocide Victims

A boy at a grave during the 2006 funeral of genocide victims

The term *Bosnian genocide* is used to refer either to the genocide committed by Serb forces in Srebrenica in 1995, or to ethnic cleansing that took place during the 1992–1995 Bosnian War.

In 2001, the International Criminal Tribunal for the Former Yugoslavia (ICTY) judged that the 1995 Srebrenica massacre was an act of genocide.

On 26 February 2007, the International Court of Justice (ICJ), in the *Bosnian Genocide Case* upheld the ICTY's earlier finding that the Srebrenica massacre in Srebrenica and Zepa constituted genocide, but found that the Serbian government had not participated in a wider genocide on the territory of Bosnia and Herzegovina during the war, as the Bosnian government had claimed.

On 12 July 2007, European Court of Human Rights when dismissing the appeal by Nikola Jorgić against his conviction for genocide by a German court (Jorgic v. Germany) noted that the German courts wider interpretation of genocide has since been rejected by international courts considering similar cases. The ECHR also noted that in the 21st century "Amongst scholars, the majority have taken the view that ethnic cleansing, in the way in which it was carried out by the Serb forces in Bosnia and Herzegovina in order to expel Muslims and Croats from their homes, did not constitute genocide. However, there are also a considerable number of scholars who have suggested that these acts did amount to genocide, and the ICTY has found in the Momcilo Krajisnik case that the actus reu, of genocide was met in Prijedor "With regard to the charge of genocide, the Chamber found that in spite of evidence of acts perpetrated in the municipalities which constituted the actus reus of genocide".

About 30 people have been indicted for participating in genocide or complicity in genocide during the early 1990s in Bosnia. To date, after several plea bargains and some convictions that were successfully challenged on appeal two men, Vujadin Popović and Ljubiša Beara, have been found guilty of committing genocide, Zdravko Tolimir has been found guilty of committing genocide and conspiracy to commit genocide, and two others, Radislav Krstić and Drago Nikolić, have been found guilty of aiding and abetting genocide. Three others have been found guilty of participating in genocides in Bosnia by German courts, one of whom Nikola Jorgić lost an appeal against his conviction in the European Court of Human Rights. A further eight men, former members of the Bosnian Serb security forces were found guilty of genocide by the State Court of Bosnia and Herzegovina.

Slobodan Milošević, as the former President of Serbia and of Yugoslavia, was the most senior political figure to stand trial at the ICTY. He died on 11 March 2006 during his trial where he was accused of genocide or complicity in genocide in territories within Bosnia and Herzegovina, so no verdict was returned. In 1995, the ICTY issued a warrant for the arrest of Bosnian Serbs Radovan Karadžić and Ratko Mladić on several charges including genocide. On 21 July 2008, Karadžić was arrested in Belgrade, and he is currently in The Hague on trial accused of genocide among other crimes. Ratko Mladić was arrested on 26 May 2011 by Serbian special police in Lazarevo, Serbia. Karadzic was convicted of ten of the eleven charges laid against him and sentenced to 40 years in prison on March 24, 2016.

International Criminal Tribunal for Rwanda (1994 to present)

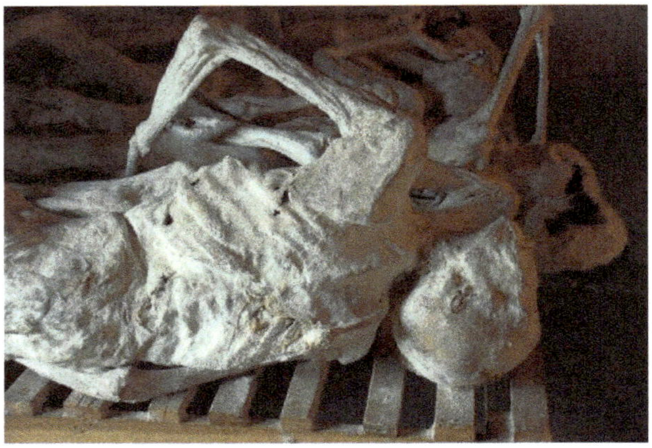

Rwandan Genocide Victims

The International Criminal Tribunal for Rwanda (ICTR) is a court under the auspices of the United Nations for the prosecution of offenses committed in Rwanda during the genocide which occurred there during April 1994, commencing on 6 April. The ICTR was created on 8 November 1994 by the Security Council of the United Nations in order to judge those people responsible for the acts of genocide and other serious violations of the international law performed in the territory of Rwanda, or by Rwandan citizens in nearby states, between 1 January and 31 December 1994.

So far, the ICTR has finished nineteen trials and convicted twenty seven accused persons. On 14 December 2009 two more men were accused and convicted for their crimes. Another twenty five persons are still on trial. Twenty-one are awaiting trial in detention, two more added on 14 December 2009. Ten are still at large. The first trial, of Jean-Paul Akayesu, began in 1997. In October 1998, Akayesu was sentenced to life imprisonment. Jean Kambanda, interim Prime Minister, pleaded guilty.

Extraordinary Chambers in the Courts of Cambodia (2003 to present)

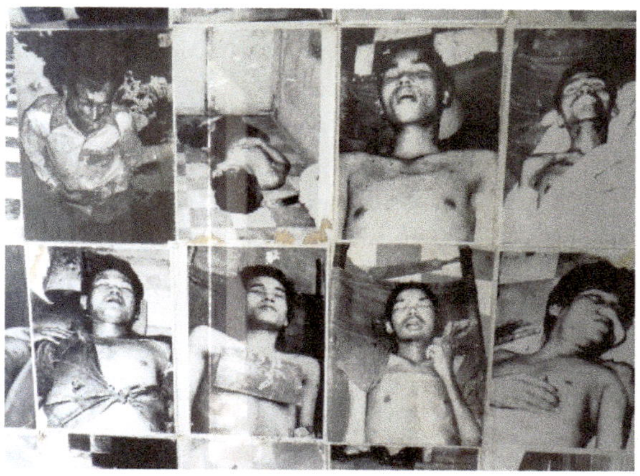

Rooms of the Tuol Sleng Genocide Museum contain thousands of photos taken by the Khmer Rouge of their victims.

Skulls in the Choeung Ek.

The Khmer Rouge, led by Pol Pot, Ta Mok and other leaders, organized the mass killing of ideologically suspect groups. The total number of victims is estimated at approximately 1.7 million Cambodians between 1975–1979, including deaths from slave labour.

On 6 June 2003 the Cambodian government and the United Nations reached an agreement to set up the Extraordinary Chambers in the Courts of Cambodia (ECCC) which would focus exclusively on crimes committed by the most senior Khmer Rouge officials during the period of Khmer Rouge rule of 1975–1979. The judges were sworn in early July 2006.

The genocide charges related to killings of Cambodia's Vietnamese and Cham minorities, which is estimated to make up tens of thousand killings and possibly more

The investigating judges were presented with the names of five possible suspects by the prosecution on 18 July 2007.

- Kang Kek Iew was formally charged with war crime and crimes against humanity and detained by the Tribunal on 31 July 2007. He was indicted on charges of war crimes and crimes against humanity on 12 August 2008. His appeal against his conviction for war crimes and crimes against humanity was rejected on 3 February 2012, and he is serving a sentence of life imprisonment.

- Nuon Chea, a former prime minister, who was indicted on charges of genocide, war crimes, crimes against humanity and several other crimes under Cambodian law on 15 September 2010. He was transferred into the custody of the ECCC on 19 September 2007. His trial, which is ongoing, started on 27 June 2011.

- Khieu Samphan, a former head of state, who was indicted on charges of genocide, war crimes, crimes against humanity and several other crimes under Cambodian law on 15 September 2010. He was transferred into the custody of the ECCC on 19 September 2007. His trial, which is ongoing, started on 27 June 2011.

- Ieng Sary, a former foreign minister, who was indicted on charges of genocide, war crimes, crimes against humanity and several other crimes under Cambodian law on 15 September 2010. He was transferred into the custody of the ECCC on 12 November 2007. His trial started on 27 June 2011, and ended with his death on 14 March 2013. He was never convicted.

- Ieng Thirith, a former minister for social affairs and wife of Ieng Sary, who was indicted on charges of genocide, war crimes, crimes against humanity and several other crimes under Cambodian law on 15 September 2010. She was transferred into the custody of the ECCC on 12 November 2007. Proceedings against her have been suspended pending a health evaluation.

There has been disagreement between some of the international jurists and the Cambodian government over whether any other people should be tried by the Tribunal.

By the International Criminal Court

Since 2002, the International Criminal Court can exercise its jurisdiction if national courts are unwilling or unable to investigate or prosecute genocide, thus being a "court of last resort," leaving the primary responsibility to exercise jurisdiction over alleged criminals to individual states. Due to the United States concerns over the ICC, the United States prefers to continue to use specially convened international tribunals for such investigations and potential prosecutions.

Darfur, Sudan

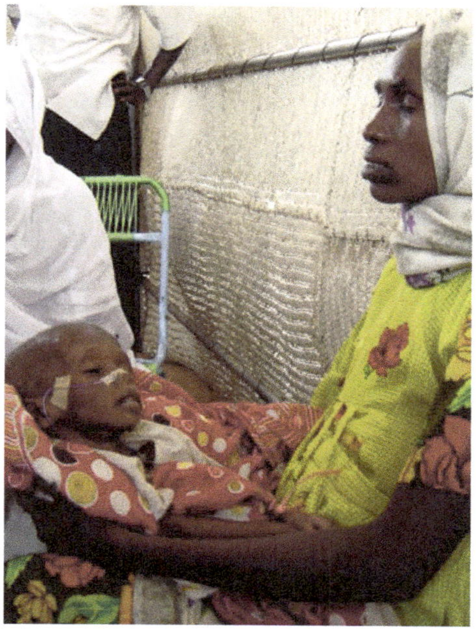

A mother with her sick baby at Abu Shouk IDP camp in North Darfur

There has been much debate over categorizing the situation in Darfur as genocide. The ongoing conflict in Darfur, Sudan, which started in 2003, was declared a "genocide" by United States Secretary of State Colin Powell on 9 September 2004 in testimony before the Senate Foreign Relations Committee. Since that time however, no other permanent member of the UN Security Council followed suit. In fact, in January 2005, an International Commission of Inquiry on Darfur, authorized by UN Security Council Resolution 1564 of 2004, issued a report to the Secretary-General stating that "the Government of the Sudan has not pursued a policy of genocide." Nevertheless, the Commission cautioned that "The conclusion that no genocidal policy has been pursued and implemented in Darfur by the Government authorities, directly or through the militias under their control, should not be taken in any way as detracting from the gravity of the crimes perpetrated in that region. International offences such as the crimes against humanity and war crimes that have been committed in Darfur may be no less serious and heinous than genocide."

In March 2005, the Security Council formally referred the situation in Darfur to the Prosecutor of the International Criminal Court, taking into account the Commission report but without mentioning any specific crimes. Two permanent members of the Security Council, the United States and China, abstained from the vote on the referral resolution. As of his fourth report to the Security Council, the Prosecutor has found "reasonable grounds to believe that the individuals identified [in the UN Security Council Resolution 1593] have committed crimes against humanity and war crimes," but did not find sufficient evidence to prosecute for genocide.

In April 2007, the Judges of the ICC issued arrest warrants against the former Minister of State for the Interior, Ahmad Harun, and a Militia Janjaweed leader, Ali Kushayb, for crimes against humanity and war crimes.

On 14 July 2008, prosecutors at the International Criminal Court (ICC), filed ten charges of war crimes against Sudan's President Omar al-Bashir: three counts of genocide, five of crimes against humanity and two of murder. The ICC's prosecutors claimed that al-Bashir "masterminded and implemented a plan to destroy in substantial part" three tribal groups in Darfur because of their ethnicity.

On 4 March 2009, the ICC issued a warrant of arrest for Omar Al Bashir, President of Sudan as the ICC Pre-Trial Chamber I concluded that his position as head of state does not grant him immunity against prosecution before the ICC. The warrant was for war crimes and crimes against humanity. It did not include the crime of genocide because the majority of the Chamber did not find that the prosecutors had provided enough evidence to include such a charge.

Genocide in history

Naked Soviet POWs held by the Nazis in Mauthausen concentration camp. "... the murder of at least 3.3 million Soviet POWs is one of the least-known of modern genocides; there is still no full-length book on the subject in English." —Adam Jones

The concept of genocide can be applied to historical events of the past. The preamble to the CPPCG states that "at all periods of history genocide has inflicted great losses on humanity."

Revisionist attempts to challenge or affirm claims of genocide are illegal in some countries. For example, several European countries ban the denial of the Holocaust or the Armenian Genocide, while in Turkey referring to the mass killings of Armenians, Greeks, Assyrians and Maronites as genocides may be prosecuted under Article 301.

William Rubinstein argues that the origin of 20th century genocides can be traced back to the collapse of the elite structure and normal modes of government in parts of Europe following the First World War:

The 'Age of Totalitarianism' included nearly all of the infamous examples of genocide in modern history, headed by the Jewish Holocaust, but also comprising the mass murders and purges of the Communist world, other mass killings carried out by Nazi Germany and its allies, and also the Armenian genocide of 1915. All these slaughters, it is argued here, had a common origin, the collapse of the elite structure and normal modes of government of much of central, eastern and southern Europe as a result of the First World War, without which surely neither Communism nor Fascism would have existed except in the minds of unknown agitators and crackpots.

— William Rubinstein, Genocide: a history

Stages of Genocide, Influences Leading to Genocide, and Efforts to Prevent it

For genocide to happen, there must be certain preconditions. Foremost among them is a national culture that does not place a high value on human life. A totalitarian society, with its assumed superior ideology, is also a precondition for genocidal acts. In addition, members of the dominant society must perceive their potential victims as less than fully human: as "pagans," "savages," "uncouth barbarians," "unbelievers," "effete degenerates," "ritual outlaws," "racial inferiors," "class antagonists," "counterrevolutionaries," and so on. In themselves, these conditions are not enough for the perpetrators to commit genocide. To do that—that is, to commit genocide—the perpetrators need a strong, centralized authority and bureaucratic organization as well as pathological individuals and criminals. Also required is a campaign of vilification and dehumanization of the victims by the perpetrators, who are usually new states or new regimes attempting to impose conformity to a new ideology and its model of society.

—M. Hassan Kakar

In 1996 Gregory Stanton, the president of Genocide Watch, presented a briefing paper called "The 8 Stages of Genocide" at the United States Department of State. In it he suggested that genocide develops in eight stages that are "predictable but not inexorable".

The Stanton paper was presented to the State Department, shortly after the Rwandan Genocide and much of its analysis is based on why that genocide occurred. The preventative measures suggested, given the briefing paper's original target audience, were those that the United States could implement directly or indirectly by using its influence on other governments.

Stage	Characteristics	Preventive measures
1. Classification	People are divided into "us and them".	"The main preventive measure at this early stage is to develop universalistic institutions that transcend... divisions."
2. Symbolization	"When combined with hatred, symbols may be forced upon unwilling members of pariah groups..."	"To combat symbolization, hate symbols can be legally forbidden as can hate speech".
3 Dehumanization	"One group denies the humanity of the other group. Members of it are equated with animals, vermin, insects, or diseases."	"Local and international leaders should condemn the use of hate speech and make it culturally unacceptable. Leaders who incite genocide should be banned from international travel and have their foreign finances frozen."
4. Organization	"Genocide is always organized... Special army units or militias are often trained and armed..."	"The U.N. should impose arms embargoes on governments and citizens of countries involved in genocidal massacres, and create commissions to investigate violations"
5. Polarization	"Hate groups broadcast polarizing propaganda..."	"Prevention may mean security protection for moderate leaders or assistance to human rights groups...Coups d'état by extremists should be opposed by international sanctions."
6. Preparation	"Victims are identified and separated out because of their ethnic or religious identity..."	"At this stage, a Genocide Emergency must be declared. ..."

7. Extermination	"It is 'extermination' to the killers because they do not believe their victims to be fully human".	"At this stage, only rapid and overwhelming armed intervention can stop genocide. Real safe areas or refugee escape corridors should be established with heavily armed international protection."
8. Denial	"The perpetrators... deny that they committed any crimes..."	"The response to denial is punishment by an international tribunal or national courts"

In April 2012, it was reported that Stanton would soon be officially adding two new stages, Discrimination and Persecution, to his original theory, which would make for a 10-stage theory of genocide.

In a paper for the Social Science Research Council Dirk Moses criticises the Stanton approach concluding:

In view of this rather poor record of ending genocide, the question needs to be asked why the "genocide studies" paradigm cannot predict and prevent genocides with any accuracy and reliability. The paradigm of "genocide studies," as currently constituted in North America in particular, has both strengths and limitations. While the moral fervor and public activism is admirable and salutary, the paradigm appears blind to its own implication in imperial projects that are themselves as much part of the problem as they are part of the solution. The US government called Darfur a genocide to appease domestic lobbies, and because the statement cost it nothing. Darfur will end when it suits the great powers that have a stake in the region.

—*Dirk Moses*

Other authors have focused on the structural conditions leading up to genocide and the psychological and social processes that create an evolution toward genocide. Ervin Staub showed that economic deterioration and political confusion and disorganization were starting points of increasing discrimination and violence in many instances of genocides and mass killing. They lead to scapegoating a group and ideologies that identified that group as an enemy. A history of devaluation of the group that becomes the victim, past violence against the group that becomes the perpetrator leading to psychological wounds, authoritarian cultures and political systems, and the passivity of internal and external witnesses (bystanders) all contribute to the probability that the violence develops into genocide. Intense conflict between groups that is unresolved, becomes intractable and violent can also lead to genocide. The conditions that lead to genocide provide guidance to early prevention, such as humanizing a devalued group, creating ideologies that embrace all groups, and activating bystander responses. There is substantial research to indicate how this can be done, but information is only slowly transformed into action.

Kjell Anderson uses a dichotomistic classification of genocides: "hot genocides, motivated by hate and the victims' threatening nature, with low-intensity cold genocides, rooted in victims' supposed inferiority."

Stampede

A stampede is uncontrolled concerted running as an act of mass impulse among herd animals or a crowd of people in which the group collectively begins running, often in an attempt to escape a perceived threat.

Wild horses stampeding

Non-human species associated with stampede behavior include cattle, elephants, blue wildebeests, walruses, wild horses, and rhinoceroses.

Cattle Stampedes

Anything unusual may start a stampede among cattle. Especially at night, things such as lighting a match, someone jumping off a horse, a horse shaking itself, a lightning strike, a tumbleweed blown into the herd, or "a horse running through a herd kicking at a saddle which has turned under its belly" have been known to cause stampedes.

A large stampede typically eliminates everything in its path. With livestock, cowboys attempt to turn the moving herd into itself, so that it runs in circles rather than running off a cliff or into a river and avoids damaging human life or property. Tactics used to make the herd turn into itself include firing a pistol, which creates noise to make the leaders of the stampede turn.

Animals that stampede, especially cattle, are less likely to do so after having eaten and spread out in smaller groups to digest. To further reduce the risk of stampedes, cowboys sometimes sing or whistle to calm the herds disquieted by nightfall. Those on watch at night avoid doing things which could startle the herd and even distance themselves before dismounting a horse or lighting a match.

Sometimes people purposefully induce cattle to stampede as a component of warfare or hunting, such as some Native Americans, who were known to cause American bison to kill themselves at a buffalo jump.

Human Stampedes and Crushes

Crushes often occur during religious pilgrimages and large entertainment events, as they tend to involve dense crowds, with people closely surrounded on all sides. Human stampedes and crushes also occur in episodes of panic (e.g. in response to a fire or explosion) as people try to get away.

Stampedes

According to experts, true "stampedes" (and "panics") rarely occur except when many people are fleeing in fear, such as from a fire, and trampling by people in such "stampede" conditions rarely causes fatal injuries.

Crushes

Crushes are very often referred to as stampedes but, unlike true stampedes, they can cause many deaths. Crowd density is more important than size. A density of four people per square meter begins to be dangerous, even if the crowd is not very large.

Academic experts who study crowd movements and crushing disasters oppose the use of the term "stampede". "The rhetoric of 'stampede' is often used to imply that the crowd is animalistic or mindless", Most reported "stampedes" are better understood as "progressive crowd collapses": beginning at densities of about six or seven persons per square meter, individuals are pressed so closely against each other they are unable to move as individuals, and shockwaves can travel through a crowd which, at such densities, behaves somewhat like a fluid. If a single person falls, or other people reach down to help, waves of bodies can be involuntarily precipitated forward into the open space. One such shockwave can create other openings in the crowd nearby, precipitating further crushing. Unable to draw breath, individuals in a crowd can also be crushed while standing. Journalistic misuse of the term "stampede", says Edwin Galea of the University of Greenwich, is the result of "pure ignorance and laziness ... it gives the impression that it was a mindless crowd only caring about themselves, and they were prepared to crush people." In reality, individuals are directly crushed by others nearby who have no choice, and those who can choose are too distant from the epicenter to be aware of what is happening.

Among causes of fatal crushes, sometimes described as "crazes", is when a large crowd is trying to get *toward* something; typically occurring when members at the back of a large crowd continue pushing forward not knowing that those at the front are being crushed, or because of something that forces them to move.

A common aftermath of a crush with serious consequences is that those responsible for the event where the crush took place, authorities such as police and government bodies, and news media blame the crowd and the victims for being out of control and causing the crush, sometimes to the extent of a full cover-up. Later analysis, sometimes after those actually responsible have retired, may show that the disaster was largely caused (in the moral and legal rather than physical sense) by actions of those planning or in authority of the event, as in the Hillsborough disaster which killed 96 football spectators; actions by the crowd were blamed until investigations three decades later found manifold errors by those responsible for organising and controlling the football event, with members of the crowd being hapless victims.

Prevention

It is believed that most major crowd disasters can be prevented by simple crowd management strategies. Human stampedes can be prevented by organization and traffic control, such as barriers. On the other hand, barriers in some cases may funnel the crowd towards an already-packed area, such as in the Hillsborough disaster. Hence barriers can be a solution in preventing or a key factor in causing a stampede. One problem is lack of feedback from people being crushed to the crowd pressing behind – feedback can instead be provided by police, organizers, or other observers, particularly raised observers, such as on platforms or horseback, who can survey the crowd and use loudspeakers to communicate and direct a crowd.

There is risk of a crush when crowd density exceeds about four people per square meter. For a person in a crowd a signal of danger, and a warning to get out of the crowd if possible, is the sensation of being touched on all four sides. A later, more serious, warning is when one feels shock waves travelling through the crowd, due to people at the back pushing forward against people at the front with nowhere to go. Keith Still of the Fire Safety Engineering Group, University of Greenwich, said "Be aware of your surroundings. Look ahead. Listen to the crowd noise. If you start finding yourself in a crowd surge, wait for the surge to come, go with it, and move sideways. Keep moving with it and sideways, with it and sideways."

After the 1883 crush known as the Victoria Hall disaster which killed 183 children, a law was passed in England which required all public entertainment venues to be equipped with doors that open outwards, for example using crash bar latches that open when pushed. Crash bars are required by various building codes.

Causes of Death

Deaths from human crushes and stampedes are found to be caused primarily by compressive asphyxiation; trampling is a lesser killer. This is due to *crowd crush* or *crowd collapse* In a crowd crush people are subjected to compressive forces by being pushed from all sides (or against a barrier such as a wall) with nowhere to move into. In a progressive crowd collapse one person falls, creating a space in the crowd into which others fall, creating an even larger hole. Those who have fallen are squashed by the weight of many people on top of them (vertical stacking). Compression in either case is often fatal. A crush is typical of a crowd pushed into a confined area; a progressive crowd collapse may occur in a large crowd moving steadily forward along a confined route.

Examples of Stampedes and Crushes

- 11 February 1823: Carnival tragedy of 1823 where about 110 boys died after falling down a flight of steps to get bread
- 16 June 1883: Victoria Hall disaster where 183 children died rushing to get treats behind a narrow door at the end of a downward staircase
- 24 December 1913: Italian Hall disaster 73 people died trying to escape from a false fire alarm at a crowded Christmas party
- 6 June 1941: during a Japanese bombing of Chongqing, China, 1000 people were killed in a stampede at the Jiaochangkou tunnel, an access point to an air raid shelter.
- 2 July 1990: 1990 Mecca tunnel tragedy: 1426 people killed
- 4 February 2006: Wowowee stampede: 73 people were killed and about 400 were injured

during a stampede outside the PhilSports Stadium in Pasig City, Philippines. About 30,000 people had been gathered outside the stadium waiting to participate in the first anniversary episode of the former television variety show Wowowee.

- 31 December 2014: 2014 Shanghai stampede: 36 people killed and 47 injured in Shanghai stampede during New Year's celebrations
- 24 September 2015: 2015 Hajj stampede At least 2,177 people crushed to death and 934 injured at the annual Hajj in Saudi Arabia

Riot

Teamsters, armed with pipes, riot in a clash with riot police in the Minneapolis Teamsters Strike of 1934.

Modern rioters typically wear face masks, scarves, and other headgear to conceal their identity and filter tear gas.

A riot is a form of civil disorder commonly characterized by a group lashing out in a violent public disturbance against authority, property or people. Riots typically involve vandalism and the destruction of property, public or private. The property targeted varies depending on the riot and the inclinations of those involved. Targets can include shops, cars, restaurants, state-owned institutions, and religious buildings.

Riots often occur in reaction to a perceived grievance or out of dissent. Historically, riots have occurred due to poor working or living conditions, governmental oppression, taxation or conscription, conflicts between ethnic groups, (race riot) or religions (sectarian violence, pogrom), the outcome of a sporting event (sports riot, football hooliganism) or frustration with legal channels through which to air grievances.

While individuals may attempt to lead or control a riot, riots typically consist of disorganized groups that are frequently "chaotic and exhibit herd behavior." However, there is a growing body of evidence to suggest that riots are not irrational, herd-like behavior, but actually follow inverted social norms.

T. S. Ashton, in his study of food riots among colliers, noted that "the turbulence of the colliers is, of course, to be accounted for by something more elementary than politics: it was the instinctive reaction of virility to hunger." Charles Wilson noted, "Spasmodic rises in food prices provoked keelmen on the Tyne to riot in 1709, tin miners to plunder granaries at Falmouth in 1727."

Today, some rioters have an improved understanding of the tactics used by police in riot situations. Manuals for successful rioting are available on the internet, with tips such as encouraging rioters to get the press involved, as there is more safety and attention with the cameras rolling. Citizens with video cameras may also have an effect on both rioters and police.

Dealing with riots is often a difficult task for police forces. They may use tear gas or CS gas to control rioters. Riot police may use less-than-lethal methods of control, such as shotguns that fire flexible baton rounds to injure or otherwise incapacitate rioters for easier arrest.

Classification

New York police violently attacking unemployed workers in Tompkins Square Park, 1874.

A police riot is a term for the disproportionate and unlawful use of force by a group of police against a group of civilians. This term is commonly used to describe a police attack on peaceful civilians, or provoking peaceful civilians into violence.

A prison riot is a large-scale, temporary act of concerted defiance or disorder by a group of prisoners against prison administrators, prison officers, or other groups of prisoners. It is often done to express a grievance, force change or attempt escape.

In a race riot, race or ethnicity is the key factor. The term had entered the English language in the

United States by the 1890s. Early use of the term referred to riots that were often a mob action by members of a majority racial group against people of other perceived races.

In a religious riot, the key factor is religion. The rioting mob targets people and properties of a specific religion, or those believed to belong to that religion.

Student riots are riots precipitated by students, often in higher education, such as a college or university. Student riots in the US and Western Europe in the 1960s and the 1970s were often political in nature. Student riots may also occur as a result of oppression of peaceful demonstration or after sporting events. Students may constitute an active political force in a given country. Such riots may occur in the context of wider political or social grievances.

Urban riots are riots in the context of urban decay, provoked by conditions such as discrimination, poverty, high unemployment, poor schools, poor healthcare, housing inadequacy and police brutality and bias. Urban riots are closely associated with race riots and police riots.

Sports riots such as the Nika riots can be sparked by the losing or winning of a specific team. Fans of the two teams may also fight. Sports riots may happen as a result of teams contending for a championship, a long series of matches, or scores that are close. Sports are the most common cause of riots in the United States, accompanying more than half of all championship games or series. Almost all sports riots occur in the winning team's city.

Food and bread riots are caused by harvest failures, incompetent food storage, hoarding, poisoning of food, or attacks by pests like locusts. When the public becomes desperate from such conditions, groups may attack shops, farms, homes, or government buildings to obtain bread or other staple foods like grain or salt, as in the 1977 Egyptian Bread Riots.

Effects

St. Augustine's Church on fire during the Philadelphia Nativist Riots in 1844

The economic and political effects of riots can be as complex as their origins. Property destruction and harm to individuals are often immediately measurable. During the 1992 Los Angeles riots, 2,383 people were injured, 8,000 were arrested, 51 were killed and over 700 businesses burned. Property damage was estimated at over $1 billion. At least ten of those killed were shot by police or National Guard forces.

Similarly, the 2005 civil unrest in France lasted over three weeks and spread to nearly 300 towns. By the end of the incident, over 10,000 vehicles were destroyed and over 300 buildings burned.

Over 2,800 suspected rioters were arrested and 126 police and firefighters were injured. Estimated damages were over €200 Million.

Many governments and political systems have fallen after riots, including:

- Tsarist Russia
- Ancien Régime
- British rule in India, when bread and salt riots hastened the withdrawal in 1947
- Governments across the Middle East and North Africa during the Arab Spring

Riot Control and Laws

Law enforcement teams deployed to control riots often wear body armor and shields, and may use tear gas

Riots are typically dealt with by the police, although methods differ from country to country. Tactics and weapons used can include attack dogs, water cannons, plastic bullets, rubber bullets, pepper spray, flexible baton rounds, and snatch squads. Many police forces have dedicated divisions to deal with public order situations. Some examples are the Territorial Support Group, Special Patrol Group, Compagnies Républicaines de Sécurité, Mobiele Eenheid, and Arrest units.

The policing of riots has been marred by incidents in which police have been accused of provoking rioting or crowd violence. While the weapons described above are officially designated as non-lethal, a number of people have died or been injured as a result of their use. For example, seventeen deaths were caused by rubber bullets in Northern Ireland over the thirty five years between 1970 and 2005.

Risk of Arrest

A high risk of being arrested is even more effective against rioting than severe punishments. As more and more people join the riot, the risk of being arrested goes down, which persuades still more people to join. This leads to a vicious circle, which is typically ended only by sufficient police or military presence to increase the risk of being arrested.

National laws

India

In India, rioting is an offense under the Indian Penal Code (IPC).

United Kingdom

England and Wales

The Brixton race riot in London, 1981

Riot is a statutory offence in England and Wales. It is created by section 1(1) of the Public Order Act 1986. Sections 1(1) to (5) of that Act read:

> (1) Where 12 or more persons who are present together use or threaten unlawful violence for a common purpose and the conduct of them (taken together) is such as would cause a person of reasonable firmness present at the scene to fear for his personal safety, each of the persons using unlawful violence for the common purpose is guilty of riot.
>
> (2) It is immaterial whether or not the 12 or more use or threaten unlawful violence simultaneously.
>
> (3) The common purpose may be inferred from conduct.
>
> (4) No person of reasonable firmness need actually be, or be likely to be, present at the scene.
>
> (5) Riot may be committed in private as well as in public places.

A single person can be liable for an offence of riot when they *use* violence, provided that it is shown there were at least twelve present *using or threatening* unlawful violence.

The word "violence" is defined by section 8. The violence can be against the person or against property. The *mens rea* is defined by section 6(1).

Indictment

R v Tyler and others, 96 Cr App R 332, Crim LR 60, CA.

Mode of Trial and Sentence

Riot is an indictable-only offence. A person convicted of riot is liable to imprisonment for any term not exceeding ten years, or to a fine, or to both.

See the following cases:

- R v Luttman Crim LR 127, CA
- R v Pilgrim, 5 Cr App R (S) 140, CA
- R v Keys, 84 Cr App R 204, 8 Cr App R (S) 444, Crim LR 207, CA
- R v Cooke, 9 Cr App R (S) 116, CA

Association Football Matches

In the case of riot connected to football hooliganism, the offender may be banned from football grounds for a set or indeterminate period of time and may be required to surrender their passport to the police for a period of time in the event of a club or international match, or international tournament, connected with the offence. This prevents travelling to the match or tournament in question. (The measures were brought in by the Football (Disorder) Act 2000 after rioting of England fans at Euro 2000.)

Compensation for Riot Damage

The Riot (Damages) Act 1886 and section 235 of the Merchant Shipping Act 1995..

Construction of "Riot" and Cognate Expressions in other Instruments

Section 10 of the Public Order Act 1986 now provides:

(1) In the Riot (Damages) Act 1886 . . . (compensation for riot damage) "riotous" and "riotously" shall be construed in accordance with section 1 above.

(2) In Schedule 1 to the Marine Insurance Act 1906 (form and rules for the construction of certain insurance policies) "rioters" in rule 8 and "riot" in rule 10 shall, in the application of the rules to any policy taking effect on or after the coming into force of this section, be construed in accordance with section 1 above unless a different intention appears.

(3) "Riot" and cognate expressions in any enactment in force before the coming into force of this section (other than the enactments mentioned in subsections (1) and (2) above) shall be construed in accordance with section 1 above if they would have been construed in accordance with the common law offence of riot apart from this Part.

(4) Subject to subsections (1) to (3) above and unless a different intention appears, nothing in this Part affects the meaning of "riot" or any cognate expression in any enactment in force, or other instrument taking effect, before the coming into force of this section.

Common Law Offence

The common law offence of riot was abolished for England and Wales on 1 April 1987.

History

In the past, the Riot Act had to be read by an official - with the wording exactly correct - before violent policing action could take place. If the group did not disperse after the Act was read, lethal force could legally be used against the crowd.

Section 515 of the Merchant Shipping Act 1894 formerly made provision for compensation for riot damage.

Northern Ireland

Riot is a serious offence for the purposes of Chapter 3 of the Criminal Justice (Northern Ireland) Order 2008.

Scotland

There is an offence under the law of Scotland which is known both as "mobbing" and "mobbing and rioting".

In July 1981, both Dundee and Edinburgh saw significant disorder as part of the events of that July, while in 1994 and in 2013, two years after the English riots of August 2011, Edinburgh saw rioting, albeit localised to one specific area and not part of any bigger 'riot wave'. Events in 1981 were very similar to those in England, although sources are severely limited. Both Niddrie and Craigmillar saw riots in the 1980s.

United States of America

The aftermath of a Washington, D.C. riot in April 1968.

Under United States federal law, a riot is defined as:

A public disturbance involving (1) an act or acts of violence by one or more persons part of an assemblage of three or more persons, which act or acts shall constitute a clear and present danger of, or shall result in, damage or injury to the property of any other person or to the person of any other individual or (2) a threat or threats of the commission of an act or acts of violence by one or more persons part of an assemblage of three or more persons having, individually or collectively, the ability of immediate execution of such threat or threats, where the performance of the threatened act or acts of violence would constitute a clear and present danger of, or would result in, damage or injury to the property of any other person or to the person of any other individual.18 U.S.C. § 2102.

As every state in the United States has its own laws (subject to the Supremacy Clause), each has its own definition of a riot. In the U.S. state of New York, the term *riot* is not defined explicitly, but under § 240.08 of the N.Y. Penal Law due to the fact there was much fighting in the streets, "A person is guilty of inciting to riot when one urges ten or more persons to engage in tumultuous and violent conduct of a kind likely to create public alarm."

Massacre

The Chios Massacre refers to a famous incident during the Greek War of Independence in 1822.

Twenty-six republicans were assassinated by Franco's Nationalists at the beginning of the Spanish Civil War, between August and September 1936. This mass grave is placed at the small town named as Estépar, in Northern Spain. The excavation occurred in July–August 2014.

The El Mozote massacre, El Salvador, 1981

A massacre is a specific incident which involves the killing of people, although not necessarily a crime against humanity.

Origin and History of Term

The first recorded use in English of the word *massacre* to label an event is Marlowe's (circa 1600), *The massacre at Paris* (a reference to the St. Bartholomew's Day massacre). The word ultimately derives from Middle Low German **matskelen* meaning *to slaughter*.

Massacre is also a verb that means to kill (people or, less commonly, animals) in numbers, especially brutally and indiscriminately. The first known use for this meaning was in 1588.

The term is also used allegorically for events that do not involve deaths, such as the Saturday Night Massacre—the dismissals and resignations of political appointees during Richard Nixon's Watergate scandal.

Definitions

Robert Melson's "basic working definition," reads, "by massacre we shall mean the intentional killing by political actors of a significant number of relatively defenseless people... the motives for massacre need not be rational in order for the killings to be intentional... Mass killings can be carried out for various reasons, including a response to false rumors... political massacre... should be distinguished from criminal or pathological mass killings... as political bodies we of course include the state and its agencies, but also nonstate actors..."

Mark Levene defines massacre historically as involving the murder of more than one individual, within an outrageous moral deficiency: "Although it is not possible to set unalterable rules about when multiple murders become massacres. Equally important is that massacres are not carried out by individuals, but by groups... the use of superior, even overwhelming force..." Levene excludes "legal, or even some quasi-legal, mass executions." He also points out that it is "...most often ... when the act is outside the normal moral bounds of the society witnessing it... In any war ... this killing is often acceptable."

References

- Arnold, Kathleen R., ed. (September 23, 2011). Anti-Immigration in the United States: A Historical Encyclopedia. II. ABC-CLIO, LLC. p. 461. ISBN 978-0-313-37521-7.
- Bruce Hoffman (1998). "Inside Terrorism". Columbia University Press. p. 32. ISBN 0-231-11468-0. Retrieved 2010-01-11. Google cached copy
- Philip P. Purpura (2007). Terrorism and homeland security: an introduction with applications. Butterworth-Heinemann. pp. 16–19. ISBN 0-7506-7843-7.
- Paula Nicolson. Domestic Violence and Psychology: A Critical Perspective. Taylor & Francis; 14 December 2010. ISBN 978-1-136-69861-3.
- Jacobson, N. & J. Gottman (1998). When Men Batter Women: New Insights into Ending Abusive Relationships. Simon & Schuster. ISBN 0-684-81447-1. OCLC 37748396.
- Sageman, Mark (2004). Understanding Terror Networks. Philadelphia, PA: U. of Pennsylvania Press. pp. 166–67. ISBN 978-0-8122-3808-2.
- Nicolas Werth, Karel Bartošek, Jean-Louis Panné, Jean-Louis Margolin, Andrzej Paczkowski, Stéphane Courtois, The Black Book of Communism: Crimes, Terror, Repression, Harvard University Press, 1999, hardcover, 858 pages, ISBN 0-674-07608-7
- Ricolfi, Luca (2005). "Palestinians 1981–2003". In Gambetta, Diego. Making Sense of Suicide Missions (1st ed.). Oxford, UK: Oxford University Press. pp. 76–130. ISBN 978-0-19-927699-8.
- Pastor, James F. (2009). Terrorism & Public Safety Policing: Implications of the Obama Presidency. New York, NY: Taylor & Francis. ISBN 978-1-4398-1580-9.
- Wilbur Redington Miller (29 June 2012). The Social History of Crime and Punishment in America: An Encyclopedia. SAGE Publications. p. 186. ISBN 0-7618-6137-8.
- David Churchman (9 May 2013). Why We Fight: The Origins, Nature, and Management of Human Conflict. University Press of America. p. 186. ISBN 0-7618-6137-8.
- Marrin, Albert (1985). The Secret Armies : Spies, Counterspies, and Saboteurs in World War II. New York: Atheneum. p. 37. ISBN 0-689-31165-6.
- LeVine, Peg (2010). Love and dread in Cambodia : weddings, births, and ritual harm under the Khmer Rouge. Singapore: NUS Press. ISBN 9789971694722.
- Robert Gellately & Ben Kiernan (2003). The Specter of Genocide: Mass Murder in Historical Perspective. Cambridge, UK: Cambridge University Press. p. 267. ISBN 0-521-52750-3.
- Staub, Ervin (31 July 1992). The Roots of Evil: The Origins of Genocide and Other Group Violence. Cambridge, UK: Cambridge University Press. p. 8. ISBN 0-521-42214-0.
- William Schabas War crimes and human rights: essays on the death penalty, justice and accountability, Cameron May 2008 ISBN 1-905017-63-4, ISBN 978-1-905017-63-8. p. 791
- Frank Chalk, Kurt Jonassohn The History and Sociology of Genocide: Analyses and Case Studies, Yale University Press, 1990, ISBN 0-300-04446-1
- Howard, Joshua (2004). Workers at War: Labor in China's Arsenals, 1937-1953. Stanford University Press. p. 128. ISBN 978-0804748964.
- Dadrian, Vahakn N. (1999). Warrant for Genocide: Key Elements of Turko-Armenian Conflict. New Brunswick: Transaction Publishers. p. 153. ISBN 1-56000-389-8.

4

Emergency Mitigation Strategies

A particular type of bunker that is used to protect the occupants from severe weather conditions is known as a storm shelter. Storm shelters are usually seen in the Midwest and southeast United States. The alternative strategies used in emergency mitigation are earthquake valve, fallout shelter, disaster recovery plan, dam safety system, community resilience etc. The topics elucidated in this section are of vital importance, and provides a better understanding of emergency mitigation strategies.

Storm Cellar

Storm cellar on the Texas plains.

A storm shelter or storm cellar is a type of underground bunker designed to protect the occupants from violent severe weather, particularly tornadoes. They are most frequently seen in the Midwest ("Tornado Alley") and Southeast United States where tornadoes are generally frequent and the low water table permits underground structures.

Average Storm Shelter

An average storm cellar for a single family is built close enough to the home to allow instant access in an emergency, but not so close that the house could tumble on the door during a storm, trapping the occupants inside. This is also the reason the main door on most storm cellars is mounted at an angle rather than perpendicular with the ground. An angled door allows for debris to blow up and over the door, or sand to slide off, without blocking it, and the angle also reduces the force necessary to open the door if rubble has piled up on top. Floor area is generally around eight by twelve feet (2.5 × 3.5 m), with an arched roof like that of a Quonset hut, but entirely underground. In most cases the entire structure is built of blocks faced with cement and rebar through the bricks for protection from the storm. Doing so makes it nearly impossible for the bricks to collapse. New ones are sometimes made of septic tanks that have been modified with a steel door and vents.

Most storm cellars are accessible by a covered stairwell, and at the opposite end of the structure there can be conduits for air that reach the surface, and perhaps a small window to serve as an emergency exit and also to provide some light. Storm cellars, when connected to the house, may potentially compromise security.

Fully enclosed underground storm shelters offer superior tornado protection to that of a traditional basement or cellar because they provide separate overhead cover without the risk of occupants being trapped or killed by collapsing rubble from above. For this reason they also provide the only reliable form of shelter against "violent" (EF4 and EF5) tornadoes which tend to rip the house from its foundation, removing the overhead cover which was protecting the occupant.

There are several different styles of storm cellars. There are the generic underground storm/tornado cellar, also called storm or tornado shelters, as well as the new above-ground safe rooms. A "cellar" is an underground unit, but for the sake of the specified use of a "storm cellar" to protect one from high-wind storms, it seems relevant to mention saferooms. There are two basic styles of underground storm cellars. One is the "hillside" or "embankment" and the other is the "flat" ground.

One other style of shelter is the under garage. While similar to other underground shelters, its main difference is that it is installed in a garage rather than outside. Having it installed in the garage allows access to it without having to go outside during a storm. It is sometimes not an option to have a shelter installed outside either due to insufficient space, or local ordinances.

Hillside/Embankment Shelters

Hillside or embankment models are usually installed in one of two ways. It can be installed in an existing hill/embankment or dirt is built up around a free standing unit forming a hill around it. The door can be set at an angle or vertically. There can be steps leading into the unit, or it can be installed to where the floor is level with the ground outside. The embankment storm cellar can be made from concrete, steel, fiberglass, or any other structurally sound material or composite and is usually installed in a hill or embankment, leaving only the door exposed. In some cases they can hold an entire neighborhood or town as with a community shelter. More often though, they are built to hold one or two families, specified as a residential shelter. All underground "storm or tornado" shelters must be properly anchored.

Above Ground Shelters

Above ground shelters are used in many areas of the country and by a wide variety of homeowners and businesses. Groundwater tables may make it impossible to install or build a shelter below ground, elderly or people with limited mobility may be unable to access a below ground shelter, or people may have significant phobias pertaining to below ground sheltering. FEMA P-320, Taking Shelter from the Storm: Building a Safe Room for Your Home or Small Business (2014) and ICC/NSSA Standard for the Design and Construction of Storm Shelters provide engineering and testing requirements to ensure that above ground shelters manufactured to the published specifications will withstand winds in excess of 250 mph (EF5 tornado). Above ground shelters may be built of different materials such as steel reinforced concrete or 1/8" 10 ga. hot rolled steel and may be installed inside a home, garage, or outbuilding, or as a stand-alone unit. These types of shelters are typically prefabricated and installed on a home site or commercial location.

Wind engineering specialists from Texas Tech University's National Wind Institute have done extensive research that concludes that sheltering in an above ground storm shelter that meets the engineering criteria outlined in FEMA Pub. 320 and 361 and ICC/NSSA Standard for the Design and Construction of Storm Shelters is as safe as seeking below ground shelter during massive EF4 and EF5 tornadoes. TTU engineer Joseph Dannemiller presented the research findings at a TEDx-TexasTechUniversity symposium in February 2014.

Flat-ground Shelters

The flat-ground cellars are designed so that the door is flat with the ground and can be made from any one of the materials previously described. This unit is put in a hole deep enough to cover the bottom section, and then the excavated dirt is filled in around the top and packed down. Storm shelters must be designed, built, tested, and installed properly for them to meet any of the U.S. FEMA-320 2008 "Taking Shelter from the Storm" 3rd Edition, FEMA-361, ICC-500 Standard, NPCTS (National Performance Criteria for Tornado Shelters), and ICC/NSSA Standard.

There is little to no policing of the storm-cellar production or installation by any government agency or association. For example, the Federal Emergency Management Agency and the National Storm Shelter Association do not enforce the manufacture or installation of any storm, hurricane, or tornado cellars. Some associations inspect one shelter for every 100 built.

Geolocation Services

Many storm shelter manufacturers include geolocation services or incorporate GPS technologies to assist in ensuring recovery from the shelter after a storm or other catastrophic event. In addition, shelter owners may opt to incorporate their own geolocation services in their shelter. An example of manufacturer-offered geolocation services is the Survive-a-Storm National Geolocator Service — shelter owners provide their shelter's GPS coordinates to an emergency response center that is linked to a nationwide severe weather notification system. If a storm occurs, then the emergency response center places a phone call to the shelter owner and then secondary contacts, lastly contacting local emergency response if unable to contact the shelter owner.

Additional Uses

Since it is functionally just an underground bunker, storm cellars can also be used as improvised bomb shelters or fallout shelters (although they are not usually dug as deeply or equipped with filtered ventilation). Since the underground construction makes them cool and dark, storm cellars on farmsteads in the Midwest are traditionally used as root cellars to store seasonal canned goods for consumption during the winter.

Earthquake Valve

An earthquake valve (or seismic valve) is an automatic method to shut off the low pressure regulated gas supply to a structure during a major earthquake and/or if a pipe is broken. These are applicable both to utility-supplied natural gas and to gas from liquefied petroleum gas (LPG). These small devices are installed on the property gas meter (usually between the utility company's metered installation and the structure piping) and are designed to instantly stop the natural gas supply in order to protect the structure if a gas leak or line break occurs during an earthquake.

Fires or explosions due to gas line breaks can be more damaging than the actual earthquake itself. Gas supply companies recommend that the gas supply be cut off immediately if there is a smell of gas after an earthquake; if nobody is in place to do this, an unattended earthquake valve will instantly cut off the gas.

Types of Valve

Two types of valve are commonly employed: one is sensitive to motion and the other to excessive gas flow. One of each type can be connected sequentially for maximum reliability.

Motion Sensing Caged Ball

A metal ball is retained away from an orifice by sitting upon a ring. Any shaking of the mechanism will cause the ball to roll off its ledge and fall down to block the orifice. It is reset using either an external magnetic device or an internal lift mechanism. If it is too sensitive, it may be triggered by normal vibrations such as passing vehicles. After a severe seismic event, gas piping may be damaged, requiring complete re-inspection for breaks or leaks.

Excessive Flow Sensor

A valve is closed when the flow exceeds a certain limit that is appropriate to the application. This will only operate when a pipe is broken and there is significant leakage. It may not operate in case of a small, though still dangerous, leak.

Fallout Shelter

A fallout shelter is an enclosed space specially designed to protect occupants from radioactive debris or fallout resulting from a nuclear explosion. Many such shelters were constructed as civil defense measures during the Cold War.

During a nuclear explosion, matter vaporized in the resulting fireball is exposed to neutrons from the explosion, absorbs them, and becomes radioactive. When this material condenses in the rain, it forms dust and light sandy materials that resembles ground pumice. The fallout emits alpha and beta particles, as well as gamma rays.

Much of this highly radioactive material falls to earth, subjecting anything within the line of sight to radiation, becoming a significant hazard. A fallout shelter is designed to allow its occupants to minimize exposure to harmful fallout until radioactivity has decayed to a safer level.

A sign pointing to an old fallout shelter in New York City.

History

Idealized American fallout shelter, around 1957.

During the Cold War, many countries built fallout shelters for high-ranking government officials and crucial military facilities, such as Project Greek Island and Cheyenne Mountain nuclear bunker in the United States and Canada's Emergency Government Headquarters. Plans were made, however, to use existing buildings with sturdy below-ground-level basements as makeshift fallout shelters. These buildings were usually placarded with the yellow and black trefoil sign.

The National Emergency Alarm Repeater (NEAR) program was developed in the United States 1956 during the Cold War to supplement the existing siren warning systems and radio broadcasts in the event of a nuclear attack. The N.E.A.R. civilian alarm device was engineered and tested but the program was not viable and was terminated in 1967. In the U.S. in September 1961, under the direction of Steuart L. Pittman, the federal government started the Community Fallout Shelter Program. A letter from President Kennedy advising the use of fallout shelters appeared in the September 1961 issue of *Life* magazine.

In November 1961, in *Fortune* magazine, an article by Gilbert Burck appeared that outlined the plans of Nelson Rockefeller, Edward Teller, Herman Kahn, and Chet Holifield for an enormous network of concrete lined underground fallout shelters throughout the United States sufficient to shelter millions of people to serve as a refuge in case of nuclear war.

Similar projects have been undertaken in Finland, which requires all buildings with area over 600 m^2 to have an NBC shelter, and Norway, which requires all buildings with an area over 1000 m^2 to have a shelter.

The former Soviet Union and other Eastern Bloc countries often designed their underground mass-transit and subway tunnels to serve as bomb and fallout shelters in the event of an attack.

Germany has protected shelters for 3% of its population, Austria for 30%, Finland for 70%, Sweden for 81% and Switzerland for 114%.

Switzerland

The Sonnenberg Tunnel, in Switzerland, was the world's largest civilian nuclear fallout shelter, designed to protect 20,000 civilians in the eventuality of war or disaster (civil defense function abandoned in 2006).

Switzerland built an extensive network of fallout shelters, not only through extra hardening of government buildings such as schools, but also through a building regulation requiring nuclear shelters in residential buildings since the 1960s (the first legal basis in this sense dates from 4 October 1963). Later, the law ensured that all residential buildings built after 1978 contained a nuclear shelter able to withstand a blast from a 12 megaton explosion at a distance of 700 metres. The *Federal Law on the Protection of the Population and Civil Protection* still requires nowadays that every inhabitant should have a place in a shelter close to where they live.

The Swiss authorities also maintain large communal shelters (such as the Sonnenberg Tunnel) stocked with over four months of food and fuel. The reference *Nuclear War Survival Skills* declared that, as of 1986, "Switzerland has the best civil defense system, one that already includes blast shelters for over 85 percent of all its citizens." As of 2006, there were about 300,000 shelters built in private homes, institutions and hospitals, as well as 5,100 public shelters for a total of 8.6 million places, a level of coverage equal to 114% of the population.

In Switzerland, most residential shelters are no longer stocked with the food and water required for prolonged habitation and a large number have been converted by the owners to other uses (e.g., wine cellars, ski rooms, gyms). But the owner still has the obligation to ensure the maintenance of the shelter.

Details of Shelter Construction

Concrete door of a public fallout shelter in Switzerland (2014).

Large fire door, sealing a fallout and air raid shelter inside the basement parking garage of a hotel in Germany.

Shielding

A basic fallout shelter consists of shields that reduce gamma ray exposure by a factor of 1000. The required shielding can be accomplished with 10 times the thickness of any quantity of material capable of cutting gamma ray exposure in half. Shields that reduce gamma ray intensity by 50% (1/2) include 1 cm (0.4 inch) of lead, 6 cm (2.4 inches) of concrete, 9 cm (3.6 inches) of packed earth or 150 m (500 ft) of air. When multiple thicknesses are built, the shielding multiplies. Thus, a practical fallout shield is ten halving-thicknesses of packed earth, reducing gamma rays by approximately 1024 times (2^{10}).

Usually, an expedient purpose-built fallout shelter is a trench; with a strong roof buried by c. 1 m (3 ft) of earth. The two ends of the trench have ramps or entrances at right angles to the trench, so that gamma rays cannot enter (they can travel only in straight lines). To make the overburden waterproof (in case of rain), a plastic sheet may be buried a few inches below the surface and held down with rocks or bricks.

Blast doors are designed to absorb the shock wave of a nuclear blast, bending and then returning to their original shape.

Climate Control

Dry earth is a reasonably good thermal insulator, and over several weeks of habitation, a shelter will become dangerously hot. The simplest form of effective fan to cool a shelter is a wide, heavy frame with flaps that swing in the shelter's doorway and can be swung from hinges on the ceiling. The flaps open in one direction and close in the other, pumping air. (This is a Kearny Air Pump, or KAP, named after the inventor, Cresson Kearny)

Unfiltered air is safe, since the most dangerous fallout has the consistency of sand or finely ground pumice. Such large particles are not easily ingested into the soft tissues of the body, so extensive filters are not required. Any exposure to fine dust is far less hazardous than exposure to the fallout outside the shelter. Dust fine enough to pass the entrance will probably pass through the shelter. Some shelters, however, incorporate NBC-filters for additional protection.

Locations

Effective public shelters can be the middle floors of some tall buildings or parking structures, or below ground level in most buildings with more than 10 floors. The thickness of the upper floors must form an effective shield, and the windows of the sheltered area must not view fall-out-covered ground that is closer than 1.5 km (1 mi). One of Switzerland's solutions is to utilise road tunnels passing through the mountains, with some of these shelters being able to protect tens of thousands.

Fallout shelters are not always underground. Above ground buildings with walls and roofs dense enough to afford a meaningful protection factor can be used as a fallout shelter.

Contents

A battery-powered radio may be helpful to get reports of fallout patterns and clearance. However, radio and other electronic equipment may be disabled by electromagnetic pulse. For example, even at the height of the cold war, EMP protection had been completed for only 125 of the approximately 2,771 radio stations in the United States Emergency Broadcast System. Also, only 110 of 3,000 existing Emergency Operating Centers had been protected against EMP effects. The Emergency Broadcast System has since been supplanted in the United States by the Emergency Alert System.

The reference *Nuclear War Survival Skills* includes the following supplies in a list of "Minimum Pre-Crisis Preparations": one or more shovels, a pick, a bow-saw with an extra blade, a hammer, and 4-mil polyethylene film (also any necessary nails, wire, etc.); a homemade shelter-ventilating pump (a KAP); large containers for water; a plastic bottle of sodium hypochlorite bleach; one or two KFMs and the knowledge to operate them; at least a 2-week supply of compact, nonperishable food; an efficient portable stove; wooden matches in a waterproof container; essential containers and utensils for storing, transporting, and cooking food; a hose-vented 5-gallon can, with heavy plastic bags for liners, for use as a toilet; tampons; insect screen and fly bait; any special medications needed by family members; pure potassium iodide, a 2-oz bottle, and a medicine dropper; a first-aid kit and a tube of antibiotic ointment; long-burning candles (with small wicks) sufficient for at least 14 nights; an oil lamp; a flashlight and extra batteries; and a transistor radio with extra batteries and a metal box to protect it from electromagnetic pulse.

Inhabitants should have water on hand, 1-2 gallons per person per day. Water stored in bulk containers requires less space than water stored in smaller bottles.

Kearny Fallout Meter

Commercially made Geiger counters are expensive and require frequent calibration. It is possible to construct an electrometer-type radiation meter called the Kearny fallout meter, which does not require batteries or professional calibration, from properly-scaled plans with just a coffee can or pail, gypsum board, monofilament fishing line, and aluminum foil. Plans are freely available in the public domain in the reference *Nuclear War Survival Skills* by Cresson Kearny.

Use

Inhabitants should plan to remain sheltered for at least two weeks (with an hour out at the end of the first week (which was once part of Swiss Zivilschutz)), then work outside for gradually increasing amounts of time, to four hours a day at three weeks. The normal work is to sweep or wash fallout into shallow trenches to decontaminate the area. They should sleep in a shelter for several months. Evacuation at three weeks is recommended by official authorities.

If available, inhabitants may take potassium iodide at the rate of 130 mg/day per adult (65 mg/day per child) as an additional measure to protect the thyroid gland from the uptake of dangerous radioactive iodine, a component of most fallout and reactor waste.

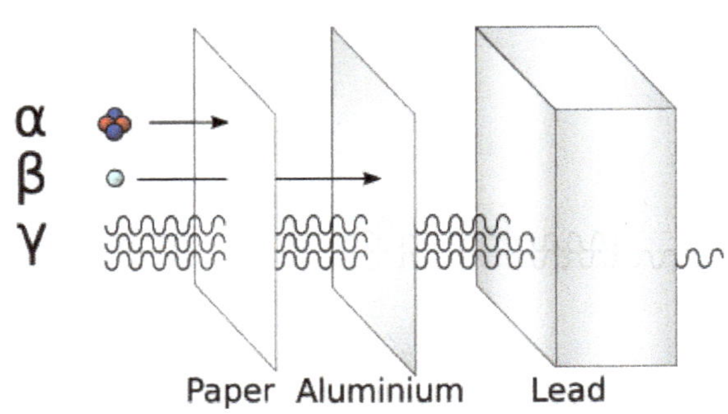

Relative abilities of three different types of ionizing radiation to penetrate solid matter.

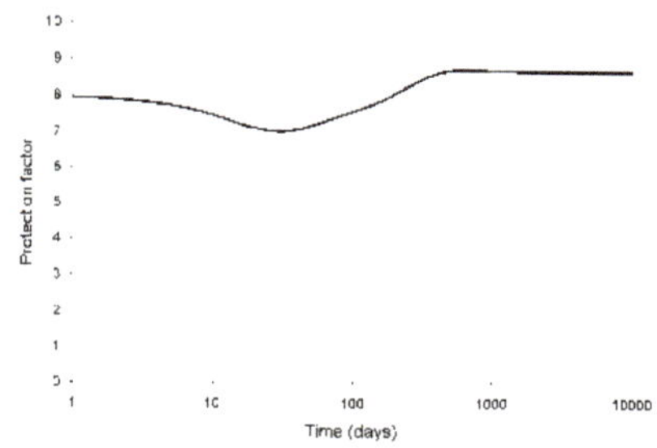

The protection factor provided by 10 cm of concrete shielding where the source is the idealised Chernobyl fallout.

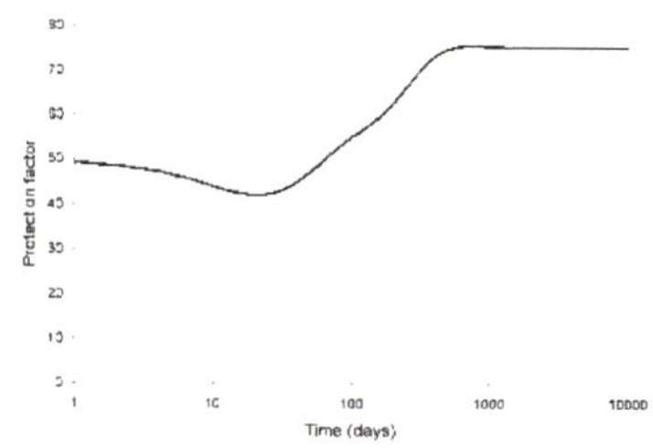

The protection factor provided by 20 cm of concrete shielding where the source is the idealised Chernobyl fallout.

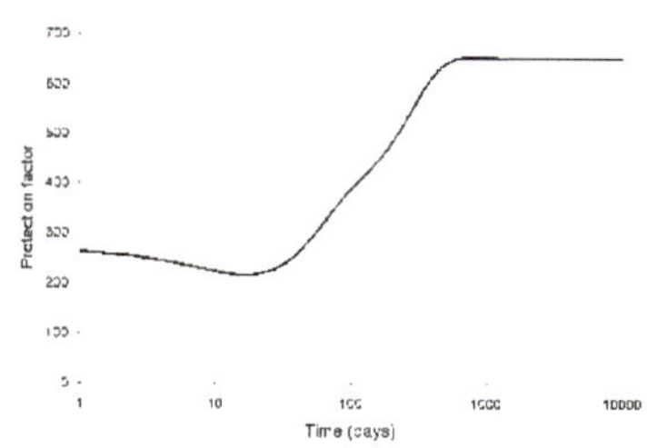

The protection factor provided by 30 cm of concrete shielding where the source is the idealised Chernobyl fallout.

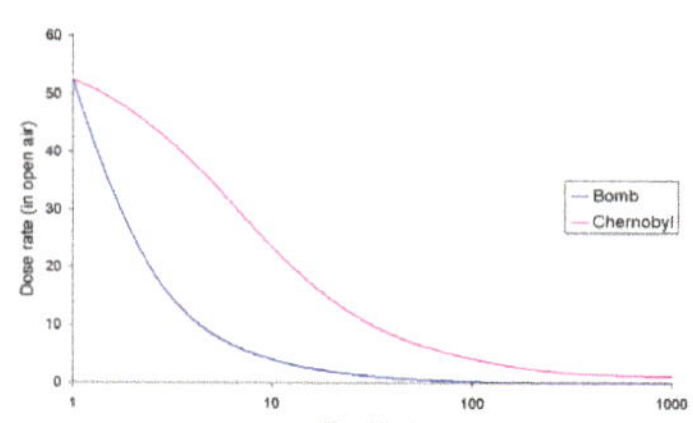

Calculated relative gamma dose rates from atomic bomb and chernobyl fallout

Different Types of Radiation Emitted by Fallout

Alpha (α)

In the vast majority of accidents, and in all atomic bomb blasts, the threat due to beta and gamma emitters is greater than that posed by the alpha emitters in the fallout. Alpha particles are identical to a helium-4 nucleus (two protons and two neutrons), and travel at speeds in excess of 5% of the speed of light. Alpha particles have little penetrating power; most cannot penetrate through human skin. Avoiding direct exposure with fallout particles will prevent injury from alpha radiation.

Beta (β)

Beta radiation consists of particles (high-speed electrons) given off by some fallout. Most beta particles cannot penetrate more than about 10 feet (3 metres) of air or about $\frac{1}{8}$ inch (3 millimetres) of water, wood, or human body tissue; or a sheet of aluminum foil. Avoiding direct exposure with fallout particles will prevent most injuries from beta radiation.

The primary dangers associated with beta radiation are internal exposure from ingested fallout particles and beta burns from fallout particles no more than a few days old. Beta burns can result from contact with highly radioactive particles on bare skin; ordinary clothing separating fresh fallout particles from the skin can provide significant shielding.

Gamma (γ)

Gamma radiation penetrates further through matter than alpha or beta radiation. Most of the design of a typical fallout shelter is intended to protect against gamma rays. Gamma rays are better absorbed by materials with high atomic numbers and high density, although neither effect is important compared to the total mass per area in the path of the gamma ray. Thus, lead is only modestly better as a gamma shield than an equal mass of another shielding material such as aluminum, concrete, water or soil.

Some gamma radiation from fallout will penetrate into even the best shelters. However, the radiation dose received while inside a shelter can be significantly reduced with proper shielding. Ten halving thicknesses of a given material can reduce gamma exposure to less than $\frac{1}{1000}$ of unshielded exposure.

Weapons Versus Nuclear Accident Fallout

The bulk of the radioactivity in nuclear accident fallout is more long-lived than that in weapons fallout. A good table of the nuclides, such as that provided by the Korean Atomic Energy Research Institute, includes the fission yields of the different nuclides. From this data it is possible to calculate the isotopic mixture in the fallout (due to fission products in bomb fallout).

Other Matters and Simple Improvements

While a person's home may not be a purpose-made shelter, it could be thought of as one if measures are taken to improve the degree of fallout protection.

Measures to Lower the Beta Dose

The main threat of beta radiation exposure comes from *hot particles* in contact with or close to the skin of a person. Also, swallowed or inhaled hot particles could cause beta burns. As it is important to avoid bringing hot particles into the shelter, one option is to remove one's outer clothing, or follow other decontamination procedures, on entry. Fallout particles will cease to be radioactive enough to cause beta burns within a few days following a nuclear explosion. The danger of gamma radiation will persist for far longer than the threat of beta burns in areas with heavy fallout exposure.

Measures to Lower the Gamma Dose Rate

The gamma dose rate due to the contamination brought into the shelter on the clothing of a person is likely to be small (by wartime standards) compared to gamma radiation that penetrates through the walls of the shelter. The following measures can be taken to reduce the amount of gamma radiation entering the shelter:

- Roofs and gutters can be cleaned to lower the dose rate in the house.
- The top inch of soil in the area near the house can be either removed or dug up and mixed with the subsoil. This reduces the dose rate as the gamma rays have to pass through the topsoil before they can irradiate anything above.
- Nearby roads can be rinsed and washed down to remove dust and debris; the fallout would collect in the sewers and gutters for easier disposal. In Kiev after the Chernobyl accident a program of road washing was used to control the spread of radioactivity.
- Windows can be bricked up, or the sill raised to reduce the hole in the shielding formed by the wall.
- Gaps in the shielding can be blocked using containers of water. While water has a much lower density than that of lead, it is still able to shield some gamma rays.
- Earth (or other dense material) can be heaped up against the exposed walls of the building; this forces the gamma rays to pass through a thicker layer of shielding before entering the house.
- Nearby trees can be removed to reduce the dose due to fallout which is on the branches and leaves. It has been suggested by the US government that a fallout shelter should not be dug close to trees for this reason.

Fallout Shelters in Popular Culture

Fallout shelter sign of the United States.

The international distinctive sign of civil defense personnel and infrastructures.

Fallout shelters feature prominently in the Robert A. Heinlein novel *Farnham's Freehold* (Heinlein built a fairly extensive shelter near his home in Colorado Springs in 1963), *Pulling Through* by Dean Ing, *A Canticle for Leibowitz* by Walter M. Miller and *Earth* by David Brin.

The *Twilight Zone* episode "The Shelter", from a Rod Serling script, deals with the consequences of actually using a shelter.

In the *Only Fools and Horses* episode "The Russians are Coming", Derek Trotter buys a lead fallout shelter, then decides to construct it in fear of an impending nuclear war caused by the Soviet Union (who were still active during the episode's creation).

In 1999 the film *Blast from the Past* was released. It is a romantic comedy film about a nuclear physicist, his wife, and son that enter a well-equipped, spacious fallout shelter during the 1962 Cuban Missile Crisis. They do not emerge until 35 years later, in 1997. The film shows their reaction to contemporary society.

The *Fallout* series of computer games depicts the remains of human civilization after an immensely destructive global nuclear war; the United States of America had built underground vaults that were advertised to protect the population against a nuclear attack, but almost all of them were in fact meant to lure subjects for long-term human experimentation.

Fallout Shelter is a free-to-play mobile simulation video game developed by Bethesda Game Studios and published by Bethesda Softworks. Part of the Fallout series, it was released for iOS devices on June 14, 2015, and is scheduled to release for Android devices in August 2015.

Paranoia, a role-playing game, takes place in a form of fallout shelter, which has become ruled by an insane computer.

The *Metro 2033* book series by Russian author Dmitry Glukhovsky depicts survivors' life in the subway systems below Moscow and Saint-Petersburg after a nuclear exchange between the Russian Federation and the United States of America.

Fallout shelters are often featured on the reality television show *Doomsday Preppers*.

The Silo series of novellas by Hugh Howey feature extensive fallout-style shelters that protect the inhabitants from an initially unknown disaster.

Blast Shelter

A blast shelter is a place where people can go to protect themselves from bomb blasts. It differs from a fallout shelter, in that its main purpose is to protect from shock waves and overpressure, instead of from radioactive precipitation, as a fallout shelter does. It is also possible for a shelter to protect from both blasts and fallout.

Blast shelters are a vital form of protection from nuclear attack and are employed in civil defense. There are above-ground, below-ground, dedicated, dual-purpose, and potential blast shelters. Dedicated blast shelters are built specifically for the purpose of blast protection. Dual-purpose blast shelters are existing structures with blast-protective properties that have been modified to accommodate people seeking protection from blasts. Potential blast shelters are existing structures or geological features exhibiting blast-protective properties that have potential to be used for protection from blasts.

Design

Blast doors in a missile control bunker at Minot Air Force Base, North Dakota.

The 25-ton blast door in the Cheyenne Mountain nuclear bunker is the main entrance to another blast door (background) beyond which the side tunnel branches into access tunnels to the main chambers.

Blast shelters deflect the blast wave from nearby explosions to prevent ear and internal injuries to people sheltering in the bunker. While frame buildings collapse from as little as 3 psi (20 kPa) of overpressure, blast shelters are regularly constructed to survive several hundred psi. This substantially decreases the likelihood that a bomb can harm the structure.

The basic plan is to provide a structure that is very strong in compression. The actual strength specification must be done individually, based on the nature and probability of the threat. A typical specification for heavy civil defence shelter in Europe during the Cold war was an overhead explosion of a 500 kiloton weapon at the height of 500 meters. Such a weapon would be used to attack soft targets (factories, administrative centres, communications) in the area.

Only the most heavy bedrock-shelters would stand a chance of surviving. However, in the countryside or in a suburb, the likely distance to the explosion is much larger, as it is improbable that anyone would waste an expensive nuclear device on such targets. The most common purpose-built structure is a steel-reinforced concrete vault or arch buried or located in the basement of a house.

Most expedient blast shelters are civil engineering structures that contain large buried tubes or pipes such as sewage or rapid transit tunnels. Even these, nonetheless, require several additions to serve properly: blast doors, air-filtration and ventilation equipment, secondary exits, and air-proofing.

Improvised purpose-built blast shelters normally use earthen arches or vaults. To form these, a narrow (1-2 metre-wide) flexible tent of thin wood is placed in a deep trench (usually the apex of the tent is below grade), and then covered with cloth or plastic, and then covered with 1–2 meters of tamped earth. Shelters of this type are approved field expedient blast shelters of both the U.S. and China. Entrances are constructed from thick wooden frames. Blast valves are to be constructed from tire-treads laid on thick wooden grids.

Nuclear bunkers must also cope with the underpressure that lasts for several seconds after the shock wave passes, and prompt radiation. The overburden and structure provide substantial radiation shielding, and the negative pressure is usually only 1/3 of the overpressure.

The doors must be at least as strong as the walls. The usual design is a trap-door, to minimize the size and expense. In dual-purpose shelters, which have a secondary peace time use, the door may be normal. To reduce the weight, the door is normally constructed of steel, with a fitted steel lintel and frame welded to the steel-reinforcement of the concrete. The shelter should be located so that there is no combustible material directly outside it.

If the door is on the surface and will be exposed to the blast wave, the edge of the door is normally counter-sunk in the frame so that the blast wave or a reflection cannot lift the edge. If possible, this should be avoided, and the door built so that it is sheltered from the blast wave by other structures. The most useful construction is to build the door behind a 90°-turn in a corridor that has an exit for the overpressure.

A bunker commonly has two doors, one of which is convenient, and in peace time use, and the other is strong. Naturally, the shelter must always have a secondary exit which can be used if the primary door is blocked by debris. Door shafts may double as ventilation shafts to reduce the digging, although this is inadvisable.

Door of a light civil defence shelter in Finland

A large ground shock can move the walls of a bunker several centimeters in a few milliseconds. Bunkers designed for large ground shocks must have sprung internal buildings, hammocks, or bean-bag chairs to protect inhabitants from the walls and floors. However, most civilian-built improvised shelters do not need these as their structure cannot stand a shock large enough to seriously damage the occupants.

Earth is an excellent insulator. In bunkers inhabited for prolonged periods, large amounts of ventilation or air-conditioning must be provided to prevent heat prostration. In bunkers designed for war-time use, manually operated ventilators must be provided because supplies of electricity or gas are unreliable. The simplest form of effective fan to cool a shelter is a wide, heavy frame with flaps that swings in the shelter's doorway and can be swung from hinges on the ceiling.

The flaps open in one direction and close in the other, pumping air. (This is a Kearny Air Pump, or KAP, named after the inventor Cresson Kearny.) Kearny asserts, based on field testing, that air filtration is not normally needed in a nuclear shelter. He asserts that fallout is either large enough to fall to the ground, or so fine that it will not settle and thus has little bulk to emit radiation. However, if possible, shelters should have air-filtration to stop chemical, biological and nuclear impurities which may abound after an explosion.

Ventilation openings in a bunker must be protected by blast valves. A blast valve is closed by a shock wave, but otherwise remains open. If the bunker is in a built-up area, it may include water-cooling or an immersion tube and breathing tubes to protect inhabitants from fire storms. In these cases, the secondary exit is also most useful.

Bunkers must also protect the inhabitants from normal weather, including rain, summer heat and winter cold. A normal form of rainproofing is to place plastic film on the bunker's main structure before burying it. Thick (5-mil or 125 μm), inexpensive polyethylene film serves quite well, because the overburden protects it from degradation by wind and sunlight. Naturally, a buried or basement-situated reinforced-concrete shelter usually has the normal appearance of a building.

When a house is purpose-built with a blast shelter, the normal location is a reinforced below-grade bathroom with large cabinets. In apartment houses, the shelter may double as storage space, as long as it can be swiftly emptied for its primary use. A shelter can easily be added in a new basement construction by taking an existing corner and adding two poured walls and a ceiling.

Some vendors provide true blast shelters engineered to provide good protection to individual families at modest cost. One common design approach uses fiber-reinforced plastic shells. Compressive protection may be provided by inexpensive earth arching. The overburden is designed to shield from radiation. To prevent the shelter from floating to the surface in high groundwater, some designs have a skirt held-down with the overburden. A properly designed, properly installed home shelter does not become a sinkhole in the lawn. In Switzerland, which requires shelters for private apartment blocks and large private houses, the lightest shelters are constructed of stainless steel.

Subways

A blast door of a subway shelter in Singapore

During World War II, people in London and Moscow survived bombing during the German Blitz by taking refuge in the undergrounds, e.g., the London underground. In the second half of the 20th century, metro stations in eastern Europe and the USSR were constructed to serve as blast shelters.

Stations of the Pyongyang Metro in North Korea, constructed 110 metres (360 ft) below ground in the 1960s and 1970s, are designed as nuclear blast shelters and each station entrance has thick steel blast doors.

Seismic Retrofit

Seismic retrofitting is the modification of existing structures to make them more resistant to seismic activity, ground motion, or soil failure due to earthquakes. With better understanding of seismic demand on structures and with our recent experiences with large earthquakes near urban centers, the need of seismic retrofitting is well acknowledged. Prior to the introduction of modern seismic codes in the late 1960s for developed countries (US, Japan etc.) and late 1970s for many other parts of the world (Turkey, China etc.), many structures were designed without adequate detailing and reinforcement for seismic protection. In view of the imminent problem, various research work has been carried out. State-of-the-art technical guidelines for seismic assessment, retrofit and rehabilitation have been published around the world - such as the ASCE-SEI 41 and the New Zealand Society for Earthquake Engineering (NZSEE)'s guidelines. These codes must be regularly updated; the 1994 Northridge earthquake brought to light the brittleness of welded steel frames, for example.

The retrofit techniques outlined here are also applicable for other natural hazards such as tropical cyclones, tornadoes, and severe winds from thunderstorms. Whilst current practice of seismic retrofitting is predominantly concerned with structural improvements to reduce the seismic hazard of using the structures, it is similarly essential to reduce the hazards and losses from non-structural elements. It is also important to keep in mind that there is no such thing as an earthquake-proof structure, although seismic performance can be greatly enhanced through proper initial design or subsequent modifications.

Infill shear trusses — University of California dormitory, Berkeley

External bracing of an existing reinforced concrete parking garage (Berkeley)

Port Authority Bus Terminal

Strategies

Seismic retrofit (or rehabilitation) strategies have been developed in the past few decades following the introduction of new seismic provisions and the availability of advanced materials (e.g. fiber-reinforced polymers (FRP), fiber reinforced concrete and high strength steel). Retrofit strategies are different from retrofit techniques, where the former is the basic approach to achieve an overall retrofit performance objective, such as increasing strength, increasing deformability, reducing deformation demands while the latter is the technical methods to achieve that strategy, for example FRP jacketing.

- Increasing the global capacity (strengthening). This is typically done by the addition of cross braces or new structural walls.
- Reduction of the seismic demand by means of supplementary damping and/or use of base isolation systems.
- Increasing the local capacity of structural elements. This strategy recognises the inherent capacity within the existing structures, and therefore adopt a more cost-effective approach to selectively upgrade local capacity (deformation/ductility, strength or stiffness) of individual structural components.
- Selective weakening retrofit. This is a counter intuitive strategy to change the inelastic mechanism of the structure, while recognising the inherent capacity of the structure.
- Allowing sliding connections such as passageway bridges to accommodate additional movement between seismically independent structures.
- Addition of seismic friction dampers to brace weak structures and provide damping.

Performance Objectives

In the past, seismic retrofit was primarily applied to achieve public safety, with engineering solutions limited by economic and political considerations. However, with the development of Performance based earthquake engineering (PBEE), several levels of performance objectives are gradually recognised:

- Public safety only. The goal is to protect human life, ensuring that the structure will not collapse upon its occupants or passersby, and that the structure can be safely exited. Under severe seismic conditions the structure may be a total economic write-off, requiring tear-down and replacement.
- Structure survivability. The goal is that the structure, while remaining safe for exit, may require extensive repair (but not replacement) before it is generally useful or considered safe for occupation. This is typically the lowest level of retrofit applied to bridges.
- Structure functionality. Primary structure undamaged and the structure is undiminished in utility for its primary application. A high level of retrofit, this ensures that any required repairs are only “cosmetic” - for example, minor cracks in plaster, drywall and stucco. This is the minimum acceptable level of retrofit for hospitals.
- Structure unaffected. This level of retrofit is preferred for historic structures of high cultural significance.

Techniques

Common seismic retrofitting techniques fall into several categories:

One of many “earthquake bolts” found throughout period houses in the city of Charleston subsequent to the Charleston earthquake of 1886. They could be tightened and loosened to support the house without having to otherwise demolish the house due to instability. The bolts were directly loosely connected to the supporting frame of the house.

External Post-tensioning

The use of external post-tensioning for new structural systems have been developed in the past decade. Under the PRESS (Precast Seismic Structural Systems), a large-scale U.S./Japan joint research program, unbonded post-tensioning high strength steel tendons have been used to achieve a moment-resisting system that has self-centering capacity. An extension of the same idea for seismic retrofitting has been experimentally tested for seismic retrofit of California bridges under a Caltrans research project and for seismic retrofit of non-ductile reinforced concrete frames. Pre-stressing can increase the capacity of structural elements such as beam, column and beam-column joints. It should be noted that external pre-stressing has been used for structural upgrade for gravity/live loading since the 1970s.

Base Isolators

Base isolation is a collection of structural elements of a building that should substantially decouple the building's structure from the shaking ground thus protecting the building's integrity and enhancing its seismic performance. This earthquake engineering technology, which is a kind of seismic vibration control, can be applied both to a newly designed building and to seismic upgrading of existing structures. Normally, excavations are made around the building and the building is separated from the foundations. Steel or reinforced concrete beams replace the connections to the foundations, while under these, the isolating pads, or base isolators, replace the material removed. While the base isolation tends to restrict transmission of the ground motion to the building, it also keeps the building positioned properly over the foundation. Careful attention to detail is required where the building interfaces with the ground, especially at entrances, stairways and ramps, to ensure sufficient relative motion of those structural elements.

Supplementary Dampers

Supplementary dampers absorb the energy of motion and convert it to heat, thus "damping" resonant effects in structures that are rigidly attached to the ground. In addition to adding energy dissipation capacity to the structure, supplementary damping can reduce the displacement and acceleration demand within the structures. In some cases, the threat of damage does not come from the initial shock itself, but rather from the periodic resonant motion of the structure that repeated ground motion induces. In the practical sense, supplementary dampers act similarly to Shock absorbers used in automotive suspensions.

Tuned Mass Dampers

Tuned mass dampers (TMD) employ movable weights on some sort of springs. These are typically employed to reduce wind sway in very tall, light buildings. Similar designs may be employed to impart earthquake resistance in eight to ten story buildings that are prone to destructive earthquake induced resonances.

Slosh Tank

A slosh tank is a large tank of fluid placed on an upper floor. During a seismic event, the fluid in this tank will slosh back and forth, but is directed by baffles - partitions that prevent the tank itself becoming resonant; through its mass the water may change or counter the resonant period of the building. Additional kinetic energy can be converted to heat by the baffles and is dissipated through the water - any temperature rise will be insignificant. One Rincon Hill in San Francisco is a recent skyscraper with a rooftop slosh tank.

Active Control System

Very tall buildings ("skyscrapers"), when built using modern lightweight materials, might sway uncomfortably (but not dangerously) in certain wind conditions. A solution to this problem is to include at some upper story a large mass, constrained, but free to move within a limited range, and moving on some sort of bearing system such as an air cushion or hydraulic film. Hydraulic pistons, powered by electric pumps and accumulators, are actively driven to counter the wind forces and natural resonances. These may also, if properly designed, be effective in controlling excessive motion - with or without applied power - in an earthquake. In general, though, modern steel frame high rise buildings are not as subject to dangerous motion as are medium rise (eight to ten story) buildings, as the resonant period of a tall and massive building is longer than the approximately one second shocks applied by an earthquake.

Adhoc Addition of Structural Support/reinforcement

The most common form of seismic retrofit to lower buildings is adding strength to the existing structure to resist seismic forces. The strengthening may be limited to connections between existing building elements or it may involve adding primary resisting elements such as walls or frames, particularly in the lower stories.

Connections between Buildings and their Expansion Additions

Frequently, building additions will not be strongly connected to the existing structure, but simply placed adjacent to it, with only minor continuity in flooring, siding, and roofing. As a result, the addition may have a different resonant period than the original structure, and they may easily detach from one another. The relative motion will then cause the two parts to collide, causing severe structural damage. Seismic modification will either tie the two building components rigidly together so that they behave as a single mass or it will employ dampers to expend the energy from relative motion, with appropriate allowance for this motion, such as increased spacing and sliding bridges between sections.

Exterior Reinforcement of Building

Exterior Concrete Columns

Historic buildings, made of unreinforced masonry, may have culturally important interior detailing or murals that should not be disturbed. In this case, the solution may be to add a number of steel, reinforced concrete, or poststressed concrete columns to the exterior. Careful attention must be paid to the connections with other members such as footings, top plates, and roof trusses.

Infill Shear Trusses

Shown here is an exterior shear reinforcement of a conventional reinforced concrete dormitory building. In this case, there was sufficient vertical strength in the building columns and sufficient shear strength in the lower stories that only limited shear reinforcement was required to make it earthquake resistant for this location near the Hayward fault.

Massive Exterior Structure

In other circumstances, far greater reinforcement is required. In the structure shown at right — a parking garage over shops — the placement, detailing, and painting of the reinforcement becomes itself an architectural embellishment.

Typical Retrofit Scenario & Solution

Soft-story Failure

Partial failure due to inadequate shear structure at garage level.
Damage in San Francisco due to the Loma Prieta event.

This collapse mode is known as *soft story collapse*. In many buildings the ground level is designed for different uses than the upper levels. Low rise residential structures may be built over a parking garage which have large doors on one side. Hotels may have a tall ground floors to allow for a grand entrance or ballrooms. Office buildings may have stores in the ground floor which desire continuous windows for display.

Traditional seismic design assumes that the lower stories of a building are stronger than the upper stories and where this is not the case—if the lower story is less strong than the upper structure—the structure will not respond to earthquakes in the expected fashion. Using modern design methods, it is possible to take a weak story into account. Several failures of this type in one large apartment complex caused most of the fatalities in the 1994 Northridge earthquake.

Typically, where this type of problem is found, the weak story is reinforced to make it stronger than the floors above by adding shear walls or moment frames. Moment frames consisting of inverted U bents are useful in preserving lower story garage access, while a lower cost solution may be to use shear walls or trusses in several locations, which partially reduce the usefulness for automobile parking but still allow the space to be used for other storage.

Beam-column Joint Connections

Corner joint steel reinforcement and high tensile strength rods with grouted anti-burst jacket below

Beam-column joint connections are a common structural weakness in dealing with seismic retrofitting. Prior to the introduction of modern seismic codes in early 1970s, beam-column joints were typically non-engineered or designed. Laboratory testings have confirmed the seismic vulnerability of these poorly detailed and under-designed connections. Failure of beam-column joint connections can typically lead to catastrophic collapse of a frame-building, as often observed in recent earthquakes

For reinforced concrete beam-column joints - various retrofit solutions have been proposed and tested in the past 20 years. Philosophically, the various seismic retrofit strategies discussed above can be implemented for reinforced concrete joints. Concrete or steel jacketing have been a popular retrofit technique until the advent of composite materials such as Carbon fiber-reinforced polymer (FRP). Composite materials such as carbon FRP and aramic FRP have been extensively tested for use in seismic retrofit with some success. One novel technique includes the use of selective weakening of the beam and added external post-tensioning to the joint in order to achieve flexural hinging in the beam, which is more desirable in terms of seismic design.

Widespread weld failures at beam-column joints of low-to-medium rise steel buildings during the Northridge 1994 earthquake for example, have shown the structural defiencies of these 'modern-designed' post-1970s welded moment-resisting connections. A subsequent SAC research project has documented, tested and proposed several retrofit solutions for these welded steel moment-resisting connections. Various retrofit solutions have been developed for these welded joints - such as a) weld strengthening and b) addition of steel haunch or 'dog-bone' shape flange.

Following the Northridge earthquake, a number of steel moment -frame buildings were found to have experienced brittle fractures of beam to column connections. Discovery of these unanticipated brittle fractures of framing connections was alarming to engineers and the building industry. Starting in the 1960s, engineers began to regard welded steel moment-frame buildings as being among the most ductile systems contained in the building code. Many engineers believed that steel moment-frame buildings were essentially invulnerable to earthquake induced damage and thought that should damage occur, it would be limited to ductile yielding of members and connections. Observation of damage sustained by buildings in the 1994 Northridge earthquake indicated that contrary to the intended behavior, in many cases, brittle fractures initiated within the connections at very low levels of plastic demand. In September, 1994, The SAC joint Venture, AISC, AISI, and NIST jointly convened an international workshop in Los Angeles to coordinate the efforts of various participants and to lay the foundation for systematic investigation and resolution of the problem. In September 1995 the SAC Joint Venture entered into a contractual agreement with FEMA to conduct Phase II of the SAC Steel project. Under Phase II, SAC continued its extensive problem-focused study of the performance of moment resisting steel frames and connections of various configurations, with the ultimate goal of developing seismic design criteria for steel construction. As a result of these studies it is now known that the typical moment-resisting connection detail employed in steel moment frame construction prior to the 1994 Northridge earthquake had a number of features that rendered it inherently susceptible to brittle fracture.

Shear Failure within Floor Diaphragm

Floors in wooden buildings are usually constructed upon relatively deep spans of wood, called joists, covered with a diagonal wood planking or plywood to form a subfloor upon which the finish floor surface is laid. In many structures these are all aligned in the same direction. To prevent the beams from tipping over onto their side, blocking is used at each end, and for additional stiffness, blocking or diagonal wood or metal bracing may be placed between beams at one or more points in their spans. At the outer edge it is typical to use a single depth of blocking and a perimeter beam overall.

If the blocking or nailing is inadequate, each beam can be laid flat by the shear forces applied to the building. In this position they lack most of their original strength and the structure may further collapse. As part of a retrofit the blocking may be doubled, especially at the outer edges of the building. It may be appropriate to add additional nails between the sill plate of the perimeter wall erected upon the floor diaphragm, although this will require exposing the sill plate by removing interior plaster or exterior siding. As the sill plate may be quite old and dry and substantial nails must be used, it may be necessary to pre-drill a hole for the nail in the old wood to avoid splitting. When the wall is opened for this purpose it may also be appropriate to tie vertical wall elements into the foundation using specialty connectors and bolts glued with epoxy cement into holes drilled in the foundation.

Sliding off Foundation and "Cripple wall" Failure

House slid off of foundation

Low cripple wall collapse and detachment of structure from concrete stairway

Single or two story wood-frame domestic structures built on a perimeter or slab foundation are relatively safe in an earthquake, but in many structures built before 1950 the sill plate that sits between the concrete foundation and the floor diaphragm (perimeter foundation) or studwall (slab foundation) may not be sufficiently bolted in. Additionally, older attachments (without substantial corrosion-proofing) may have corroded to a point of weakness. A sideways shock can slide the building entirely off of the foundations or slab.

Often such buildings, especially if constructed on a moderate slope, are erected on a platform connected to a perimeter foundation through low stud-walls called "cripple wall" or *pin-up*. This low wall structure itself may fail in shear or in its connections to itself at the corners, leading to the building moving diagonally and collapsing the low walls. The likelihood of failure of the pin-up can be reduced by ensuring that the corners are well reinforced in shear and that the shear panels are well connected to each other through the corner posts. This requires structural grade sheet plywood, often treated for rot resistance. This grade of plywood is made without interior unfilled knots and with more, thinner layers than common plywood. New buildings designed to resist earthquakes will typically use OSB (oriented strand board), sometimes with metal joins between panels, and with well attached stucco covering to enhance its performance. In many modern tract homes, especially those built upon expansive (clay) soil the building is constructed upon a single and relatively thick monolithic slab, kept in one piece by high tensile rods that are stressed after the slab has set. This poststressing places the concrete under compression - a condition under which it is extremely strong in bending and so will not crack under adverse soil conditions.

Multiple Piers in Shallow Pits

Some older low-cost structures are elevated on tapered concrete pylons set into shallow pits, a method frequently used to attach outdoor decks to existing buildings. This is seen in conditions of damp soil, especially in tropical conditions, as it leaves a dry ventilated space under the house, and in far northern conditions of permafrost (frozen mud) as it keeps the building's warmth from destabilizing the ground beneath. During an earthquake, the pylons may tip, spilling the building to the ground. This can be overcome by using deep-bored holes to contain cast-in-place reinforced pylons, which are then secured to the floor panel at the corners of the building. Another technique is to add sufficient diagonal bracing or sections of concrete shear wall between pylons.

Reinforced Concrete Column Burst

Jacketed and grouted column on left, unmodified on right

Reinforced concrete columns typically contain large diameter vertical rebar (reinforcing bars) arranged in a ring, surrounded by lighter-gauge hoops of rebar. Upon analysis of failures due to earthquakes, it has been realized that the weakness was not in the vertical bars, but rather in inadequate strength and quantity of hoops. Once the integrity of the hoops is breached, the vertical rebar can flex outward, stressing the central column of concrete. The concrete then simply crumbles into small pieces, now unconstrained by the surrounding rebar. In new construction a greater amount of hoop-like structures are used.

One simple retrofit is to surround the column with a jacket of steel plates formed and welded into a single cylinder. The space between the jacket and the column is then filled with concrete, a process called grouting. Where soil or structure conditions require such additional modification, additional pilings may be driven near the column base and concrete pads linking the pilings to the pylon are fabricated at or below ground level. In the example shown not all columns needed to be modified to gain sufficient seismic resistance for the conditions expected. (This location is about a mile from the Hayward Fault Zone.)

Reinforced Concrete Wall Burst

Concrete walls are often used at the transition between elevated road fill and overpass structures. The wall is used both to retain the soil and so enable the use of a shorter span and also to transfer the weight of the span directly downward to footings in undisturbed soil. If these walls are inadequate they may crumble under the stress of an earthquake's induced ground motion.

One form of retrofit is to drill numerous holes into the surface of the wall, and secure short L-shaped sections of rebar to the surface of each hole with epoxy adhesive. Additional vertical and horizontal rebar is then secured to the new elements, a form is erected, and an additional layer of concrete is poured. This modification may be combined with additional footings in excavated trenches and additional support ledgers and tie-backs to retain the span on the bounding walls.

Brick wall Resin and Glass Fiber Reinforcement

Brick building structures have been reinforced with coatings of glass fiber and appropriate resin (epoxy or polyester). In lower floors these may be applied over entire exposed surfaces, while in upper floors this may be confined to narrow areas around window and door openings. This application provides tensile strength that stiffens the wall against bending away from the side with the application. The efficient protection of an entire building requires extensive analysis and engineering to determine the appropriate locations to be treated.

Lift

Where moist or poorly consolidated alluvial soil interfaces in a "beach like" structure against underlying firm material, seismic waves traveling through the alluvium can be amplified, just as are water waves against a sloping beach. In these special conditions, vertical accelerations up to twice the force of gravity have been measured. If a building is not secured to a well-embedded foundation it is possible for the building to be thrust from (or with) its foundations into the air, usually with severe damage upon landing. Even if it is well-founded, higher portions such as upper stories or roof structures or attached structures such as canopies and porches may become detached from the primary structure.

Good practices in modern, earthquake-resistant structures dictate that there be good vertical connections throughout every component of the building, from undisturbed or engineered earth to foundation to sill plate to vertical studs to plate cap through each floor and continuing to the roof structure. Above the foundation and sill plate the connections are typically made using steel strap or sheet stampings, nailed to wood members using special hardened high-shear strength nails, and heavy angle stampings secured with through bolts, using large washers to prevent pull-through. Where inadequate bolts are provided between the sill plates and a foundation in existing construction (or are not trusted due to possible corrosion), special clamp plates may be added, each of which is secured to the foundation using expansion bolts inserted into holes drilled in an exposed face of concrete. Other members must then be secured to the sill plates with additional fittings.

Soil

One of the most difficult retrofits is that required to prevent damage due to soil failure. Soil failure can occur on a slope, a slope failure or landslide, or in a flat area due to liquefaction of water-saturated sand and/or mud. Generally, deep pilings must be driven into stable soil (typically hard mud or sand) or to underlying bedrock or the slope must be stabilized. For buildings built atop previous landslides the practicality of retrofit may be limited by economic factors, as it is not practical to stabilize a large, deep landslide. The likelihood of landslide or soil failure may also depend upon seasonal factors, as the soil may be more stable at the beginning of a wet season than at the beginning of the dry season. Such a "two season" *Mediterranean climate* is seen throughout California.

In some cases, the best that can be done is to reduce the entrance of water runoff from higher, stable elevations by capturing and bypassing through channels or pipes, and to drain water infiltrated directly and from subsurface springs by inserting horizontal perforated tubes. There are numerous locations in California where extensive developments have been built atop archaic landslides, which have not moved in historic times but which (if both water-saturated and shaken by an earthquake) have a high probability of moving *en masse*, carrying entire sections of suburban development to new locations. While the most modern of house structures (well tied to monolithic concrete foundation slabs reinforced with post tensioning cables) may survive such movement largely intact, the building will no longer be in its proper location.

Utility Pipes and Cables: Risks

Natural gas and propane supply pipes to structures often prove especially dangerous during and after earthquakes. Should a building move from its foundation or fall due to cripple wall collapse, the ductile iron pipes transporting the gas within the structure may be broken, typically at the location of threaded joints. The gas may then still be provided to the pressure regulator from higher pressure lines and so continue to flow in substantial quantities; it may then be ignited by a nearby source such as a lit pilot light or arcing electrical connection.

There are two primary methods of automatically restraining the flow of gas after an earthquake, installed on the low pressure side of the regulator, and usually downstream of the gas meter.

- A caged metal ball may be arranged at the edge of an orifice. Upon seismic shock, the ball will roll into the orifice, sealing it to prevent gas flow. The ball may later be reset by the use of an external magnet. This device will respond only to ground motion.
- A flow-sensitive device may be used to close a valve if the flow of gas exceeds a set threshold (very much like an electrical circuit breaker). This device will operate independently of seismic motion, but will not respond to minor leaks which may be caused by an earthquake.

It appears that the most secure configuration would be to use one of each of these devices in series.

Tunnels

Unless the tunnel penetrates a fault likely to slip, the greatest danger to tunnels is a landslide blocking an entrance. Additional protection around the entrance may be applied to divert any falling material (similar as is done to divert snow avalanches) or the slope above the tunnel may be stabilized in some way. Where only small- to medium-sized rocks and boulders are expected to fall, the entire slope may be covered with wire mesh, pinned down to the slope with metal rods. This is also a common modification to highway cuts where appropriate conditions exist.

Underwater Tubes

The safety of underwater tubes is highly dependent upon the soil conditions through which the tunnel was constructed, the materials and reinforcements used, and the maximum predicted earthquake expected, and other factors, some of which may remain unknown under current knowledge.

BART Tube

A tube of particular structural, seismic, economic, and political interest is the BART (Bay Area Rapid Transit) transbay tube. This tube was constructed at the bottom of San Francisco Bay through an innovative process. Rather than pushing a shield through the soft bay mud, the tube was constructed on land in sections. Each section consisted of two inner train tunnels of circular cross section, a central access tunnel of rectangular cross section, and an outer oval shell encompassing the three inner tubes. The intervening space was filled with concrete. At the bottom of the bay a trench was excavated and a flat bed of crushed stone prepared to receive the tube sections. The sections were then floated into place and sunk, then joined with bolted connections to previously-placed sections. An overfill was then placed atop the tube to hold it down. Once completed from San Francisco to Oakland, the tracks and electrical components were installed. The predicted response of the tube during a major earthquake was likened to be as that of a string of (cooked) spaghetti in a bowl of gelatin dessert. To avoid overstressing the tube due to differential movements at each end, a sliding slip joint was included at the San Francisco terminus under the landmark Ferry Building.

The engineers of the construction consortium PBTB (Parsons Brinckerhoff-Tudor-Bechtel) used the best estimates of ground motion available at the time, now known to be insufficient given modern computational analysis methods and geotechnical knowledge. Unexpected settlement of the tube has reduced the amount of slip that can be accommodated without failure. These factors have resulted in the slip joint being designed too short to ensure survival of the tube under possible (perhaps even likely) large earthquakes in the region. To correct this deficiency the slip joint must be extended to allow for additional movement, a modification expected to be both expensive and technically and logistically difficult. Other retrofits to the BART tube include vibratory consolidation of the tube's overfill to avoid potential liquefying of the overfill, which has now been completed. (Should the overfill fail there is a danger of portions of the tube rising from the bottom, an event which could potentially cause failure of the section connections.)

Bridge Retrofit

Bridges have several failure modes.

Expansion Rockers

Many short bridge spans are statically anchored at one end and attached to rockers at the other. This rocker gives vertical and transverse support while allowing the bridge span to expand and contract with temperature changes. The change in the length of the span is accommodated over a gap in the roadway by comb-like expansion joints. During severe ground motion, the rockers may jump from their tracks or be moved beyond their design limits, causing the bridge to unship from its resting point and then either become misaligned or fail completely. Motion can be constrained by adding ductile or high-strength steel restraints that are friction-clamped to beams and designed to slide under extreme stress while still limiting the motion relative to the anchorage.

Deck Rigidity

Additional diagonals were inserted under both decks of this bridge

Suspension bridges may respond to earthquakes with a side-to-side motion exceeding that which was designed for wind gust response. Such motion can cause fragmentation of the road surface, damage to bearings, and plastic deformation or breakage of components. Devices such as hydraulic dampers or clamped sliding connections and additional diagonal reenforcement may be added.

Lattice Girders, Beams, and Ties

Obsolete riveted lattice members

Lattice girders consist of two "I"-beams connected with a criss-cross lattice of flat strap or angle stock. These can be greatly strengthened by replacing the open lattice with plate members. This is usually done in concert with the replacement of hot rivets with bolts.

Bolted plate lattice replacement, forming box members

Hot Rivets

Many older structures were fabricated by inserting red-hot rivets into pre-drilled holes; the soft rivets are then peened using an air hammer on one side and a bucking bar on the head end. As these cool slowly, they are left in an annealed (soft) condition, while the plate, having been hot rolled and quenched during manufacture, remains relatively hard. Under extreme stress the hard plates can shear the soft rivets, resulting in failure of the joint.

The solution is to burn out each rivet with an oxygen torch. The hole is then prepared to a precise diameter with a reamer. A special *locator bolt*, consisting of a head, a shaft matching the reamed hole, and a threaded end is inserted and retained with a nut, then tightened with a wrench. As the bolt has been formed from an appropriate high-strength alloy and has also been heat-treated, it is not subject to either the plastic shear failure typical of hot rivets nor the brittle fracture of ordinary bolts. Any partial failure will be in the plastic flow of the metal secured by the bolt; with proper engineering any such failure should be non-catastrophic.

Fill and Overpass

Elevated roadways are typically built on sections of elevated earth fill connected with bridge-like segments, often supported with vertical columns. If the soil fails where a bridge terminates, the bridge may become disconnected from the rest of the roadway and break away. The retrofit for this is to add additional reinforcement to any supporting wall, or to add deep caissons adjacent to the edge at each end and connect them with a supporting beam under the bridge.

Another failure occurs when the fill at each end moves (through resonant effects) in bulk, in opposite directions. If there is an insufficient founding shelf for the overpass, then it may fall. Additional shelf and ductile stays may be added to attach the overpass to the footings at one or both ends. The stays, rather than being fixed to the beams, may instead be clamped to them. Under moderate loading, these keep the overpass centered in the gap so that it is less likely to slide off its founding shelf at one end. The ability for the fixed ends to slide, rather than break, will prevent the complete drop of the structure if it should fail to remain on the footings.

Viaducts

Large sections of roadway may consist entirely of viaduct, sections with no connection to the earth other than through vertical columns. When concrete columns are used, the detailing is critical. Typical failure may be in the toppling of a row of columns due either to soil connection failure or to insufficient cylindrical wrapping with rebar. Both failures were seen in the 1995 Great Hanshin earthquake in Kobe, Japan, where an entire viaduct, centrally supported by a single row of large columns, was laid down to one side. Such columns are reinforced by excavating to the foundation pad, driving additional pilings, and adding a new, larger pad, well connected with rebar alongside or into the column. A column with insufficient wrapping bar, which is prone to burst and then hinge at the bursting point, may be completely encased in a circular or elliptical jacket of welded steel sheet and grouted as described above.

Cypress Freeway viaduct collapse. Note failure of inadequate anti-burst wrapping and lack of connection between upper and lower vertical elements.

Sometimes viaducts may fail in the connections between components. This was seen in the failure of the Cypress Freeway in Oakland, California, during the Loma Prieta earthquake. This viaduct was a two-level structure, and the upper portions of the columns were not well connected to the lower portions that supported the lower level; this caused the upper deck to collapse upon the lower deck. Weak connections such as these require additional external jacketing - either through external steel components or by a complete jacket of reinforced concrete, often using stub connections that are glued (using epoxy adhesive) into numerous drilled holes. These stubs are then connected to additional wrappings, external forms (which may be temporary or permanent) are erected, and additional concrete is poured into the space. Large connected structures similar to the Cypress Viaduct must also be properly analyzed in their entirety using dynamic computer simulations.

Residential Retrofit

Side-to-side forces cause most earthquake damage. Bolting of the mudsill to the foundation and application of plywood to cripple walls are a few basic retrofit techniques which homeowners may apply to wood-framed residential structures to mitigate the effects of seismic activity. The City of San Leandro created guidelines for these procedures, as outlined in the following pamphlet. Public awareness and initiative are critical to the retrofit and preservation of existing building stock, and such efforts as those of the Association of Bay Area Governments are instrumental in providing informational resources to seismically active communities.

Wood Frame Structure

Most houses in North America are wood-framed structures. Wood is one of the best materials for earthquake-resistant construction since it is lightweight and more flexible than masonry. It is easy to work with and less expensive than steel, masonry, or concrete. In older homes the most significant weaknesses are the connection from the wood-framed walls to the foundation and the relatively weak "cripple-walls." (Cripple walls are the short wood walls that extend from the top of the foundation to the lowest floor level in houses that have raised floors.) Adding connections from the base of the wood-framed structure to the foundation is almost always an important part of a seismic retrofit. Bracing the cripple-walls to resist side-to-side forces is essential in houses with cripple walls; bracing is usually done with plywood. Oriented strand board (OSB) does not perform as consistently as plywood, and is not the favored choice of retrofit designers or installers.

Retrofit methods in older woodframe structures may consist of the following, and other methods not described here.

- The lowest plate rails of walls (usually called "mudsills" or "foundation sills" in North America) are bolted to a continuous foundation, or secured with rigid metal connectors bolted to the foundation so as to resist side-to-side forces.
- *Cripple walls* are braced with plywood.
- Selected vertical elements (typically the posts at the ends of plywood wall bracing panels) are connected to the foundation. These connections are intended to prevent the braced walls from rocking up and down when subjected to back-and-forth forces at the top of the braced walls, not to resist the wall or house "jumping" off the foundation (which almost never occurs).
- In two story buildings using "platform framing" (sometimes called "western" style construction, where walls are progressively erected upon the lower story's upper diaphragm, unlike "eastern" or *balloon framing*), the upper walls are connected to the lower walls with tension elements. In some cases, connections may be extended vertically to include retention of certain roof elements. This sort of strengthening is usually very costly with respect to the strength gained.
- Vertical posts are secured to the beams or other members they support. This is particularly important where loss of support would lead to collapse of a segment of a building. Connections from posts to beams cannot resist appreciable side-to-side forces; it is much more

important to strengthen around the perimeter of a building (bracing the cripple-walls and supplementing foundation-to-wood-framing connections) than it is to reinforce post-to-beam connections.

Wooden framing is efficient when combined with masonry, if the structure is properly designed. In Turkey, the traditional houses (bagdadi) are made with this technology. In El Salvador, wood and bamboo are used for residential construction.

Reinforced and Unreinforced Masonry

In many parts of developing countries such as Pakistan, Iran and China, unreinforced or in some cases reinforced masonry is the predominantly form of structures for rural residential and dwelling. Masonry was also a common construction form in the early part of the 20th century, which implies that a substantial number of these at-risk masonry structures would have significant heritage value. Masonry walls that are not reinforced are especially hazardous. Such structures may be more appropriate for replacement than retrofit, but if the walls are the principal load bearing elements in structures of modest size they may be appropriately reinforced. It is especially important that floor and ceiling beams be securely attached to the walls. Additional vertical supports in the form of steel or reinforced concrete may be added.

In the western United States, much of what is seen as masonry is actually brick or stone veneer. Current construction rules dictate the amount of *tie–back* required, which consist of metal straps secured to vertical structural elements. These straps extend into mortar courses, securing the veneer to the primary structure. Older structures may not secure this sufficiently for seismic safety. A weakly secured veneer in a house interior (sometimes used to face a fireplace from floor to ceiling) can be especially dangerous to occupants. Older masonry chimneys are also dangerous if they have substantial vertical extension above the roof. These are prone to breakage at the roofline and may fall into the house in a single large piece. For retrofit, additional supports may be added; however, it is extremely expensive to strengthen an existing masonry chimney to conform with contemporary design standards. It is best to simply remove the extension and replace it with lighter materials, with special metal flue replacing the flue tile and a wood structure replacing the masonry. This may be matched against existing brickwork by using very thin veneer (similar to a tile, but with the appearance of a brick).

Disaster Recovery Plan

A disaster recovery plan (DRP) is a documented process or set of procedures to recover and protect a business IT infrastructure in the event of a disaster. Such a plan, ordinarily documented in written form, specifies procedures an organization is to follow in the event of a disaster. It is "a comprehensive statement of consistent actions to be taken before, during and after a disaster." The disaster could be natural, environmental or man-made. Man-made disasters could be intentional (for example, an act of a terrorist) or unintentional (that is, accidental, such as the breakage of a man-made dam).

Given organizations' increasing dependency on information technology to run their operations, a disaster recovery plan, sometimes erroneously called a Continuity of Operations Plan (COOP), is increasingly associated with the recovery of information technology data, assets, and facilities.

Objectives

Organizations cannot always avoid disasters, but with careful planning the effects of a disaster can be minimized. The objective of a disaster recovery plan is to minimize downtime and data loss. The primary objective is to protect the organization in the event that all or part of its operations and/or computer services are rendered unusable. The plan minimizes the disruption of operations and ensures that some level of organizational stability and an orderly recovery after a disaster will prevail. Minimizing downtime and data loss is measured in terms of two concepts: the recovery time objective (RTO) and the recovery point objective (RPO).

The recovery time objective is the time within which a business process must be restored, after a major incident (MI) has occurred, in order to avoid unacceptable consequences associated with a break in business continuity. The recovery point objective (RPO) is the age of files that must be recovered from backup storage for normal operations to resume if a computer, system, or network goes down as a result of a MI. The RPO is expressed backwards in time (that is, into the past) starting from the instant at which the MI occurs, and can be specified in seconds, minutes, hours, or days. The recovery point objective (RPO) is thus the maximum acceptable amount of data loss measured in time. It is the age of the files or data in backup storage required to resume normal operations after the MI.

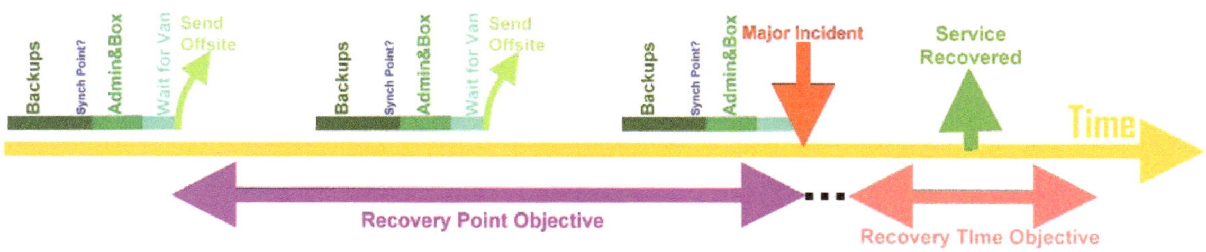

A DR plan illustrating the chronology of the **RPO** and the **RTO** with respect to the **MI**.

Relationship to the Business Continuity Plan

According to the SANS institute, the Business Continuity Plan (BCP) is a comprehensive organizational plan that includes the disaster recovery plan. The Institute further states that a Business Continuity Plan (BCP) consists of the five component plans:

- Business Resumption Plan
- Occupant Emergency Plan
- Continuity of Operations Plan
- Incident Management Plan
- Disaster Recovery Plan

The Institute states that the first three plans (Business Resumption, Occupant Emergency, and Continuity of Operations Plans) do not deal with the IT infrastructure. They further state that the Incident Management Plan (IMP) does deal with the IT infrastructure, but since it establishes structure and procedures to address cyber attacks against an organization's IT systems, it generally does not represent an agent for activating the Disaster Recovery Plan, leaving The Disaster Recovery Plan as the only BCP component of interest to IT.

Disaster Recovery Institute International states that disaster recovery is the area of business continuity that deals with *technology* recovery as opposed to the recovery of business operations.

Benefits

Like every insurance plan, there are benefits that can be obtained from the drafting of a disaster recovery plan. Some of these benefits are:

- Providing a sense of security
- Minimizing risk of delays
- Guaranteeing the reliability of standby systems
- Providing a standard for testing the plan
- Minimizing decision-making during a disaster
- Reducing potential legal liabilities
- Lowering unnecessarily stressful work environment

Types of Plans

There is no one right type of disaster recovery plan, nor is there a one-size-fits-all disaster recovery plan. However, there are three basic strategies that feature in all disaster recovery plans: (1) preventive measures, (2) detective measures, and (3) corrective measures. Preventive measures will try to prevent a disaster from occurring. These measures seek to identify and reduce risks. They are designed to mitigate or prevent an event from happening. These measures may include keeping data backed up and off site, using surge protectors, installing generators and conducting routine inspections. Detective measures are taken to discover the presence of any unwanted events within the IT infrastructure. Their aim is to uncover new potential threats. They may detect or uncover unwanted events. These measures include installing fire alarms, using up-to-date antivirus software, holding employee training sessions, and installing server and network monitoring software. Corrective measures are aimed to restore a system after a disaster or otherwise unwanted event takes place. These measures focus on fixing or restoring the systems after a disaster. Corrective measures may include keeping critical documents in the Disaster Recovery Plan or securing proper insurance policies, after a "lessons learned" brainstorming session.

A disaster recovery plan must answer at least three basic questions: (1) what is its objective and purpose, (2) who will be the people or teams who will be responsible in case any disruptions happen, and (3) what will these people do (the procedures to be followed) when the disaster strikes.

Types of Disasters

The tsunami that affected Japan in 2011, a type of natural disaster

Disasters can be natural or man-made. Man-made disasters could be intentional (for example, sabotage or an act of terrorism) or unintentional (that is, accidental, such as the breakage of a man-made dam). Disasters may encompass more than weather. They may involve Internet threats or take on other man-made manifestations such as theft.

Natural Disaster

A natural disaster is a major adverse event resulting from the earth's natural hazards. Examples of natural disasters are floods, tsunamis, tornadoes, hurricanes/cyclones, volcanic eruptions, earthquakes, heat waves, and landslides. Other types of disasters include the more cosmic scenario of an asteroid hitting the Earth.

Man-made Disasters

Man-made disasters are the consequence of technological or human hazards. Examples include stampedes, urban fires, industrial accidents, oil spills, nuclear explosions/nuclear radiation and acts of war. Other types of man-made disasters include the more cosmic scenarios of catastrophic global warming, nuclear war, and bioterrorism.

The following table categorizes some disasters and notes first response initiatives. Note that whereas the sources of a disaster may be natural (for example, heavy rains) or man-made (for example, a broken dam), the results may be similar (flooding).

	Disaster		
	Example	**Profile**	**First Response**
	Avalanche	The sudden, drastic flow of snow down a slope, occurring when either natural triggers, such as loading from new snow or rain, or artificial triggers, such as explosives or backcountry skiers, overload the snowpack	Shut off utilities; Evacuate building if necessary; Determine impact on the equipment and facilities and any disruption

Natural	Blizzard	A severe snowstorm characterized by very strong winds and low temperatures	Power off all equipment; listen to blizzard advisories; Evacuate area, if unsafe; Assess damage
	Earthquake	The shaking of the earth's crust, caused by underground volcanic forces of breaking and shifting rock beneath the earth's surface	Shut off utilities; Evacuate building if necessary; Determine impact on the equipment and facilities and any disruption
	Fire (wild)	Fires that originate in uninhabited areas and which pose the risk to spread to inhabited areas	Attempt to suppress fire in early stages; Evacuate personnel on alarm, as necessary; Notify fire department; Shut off utilities; Monitor weather advisories
	Flood	Flash flooding: Small creeks, gullies, dry streambeds, ravines, culverts or even low-lying areas flood quickly	Monitor flood advisories; Determine flood potential to facilities; Pre-stage emergency power generating equipment; Assess damage
	Freezing Rain	Rain occurring when outside surface temperature is below freezing	Monitor weather advisories; Notify employees of business closure; home; Arrange for snow and ice removal
	Heat wave	A prolonged period of excessively hot weather relative to the usual weather pattern of an area and relative to normal temperatures for the season	Listen to weather advisories; Power-off all servers after a graceful shutdown if there is imminent potential of power failure; Shut down main electric circuit usually located in the basement or the first floor
	Hurricane	Heavy rains and high winds	Power off all equipment; listen to hurricane advisories; Evacuate area, if flooding is possible; Check gas, water and electrical lines for damage; Do not use telephones, in the event of severe lightning; Assess damage
	Landslide	Geological phenomenon which includes a range of ground movement, such as rock falls, deep failure of slopes and shallow debris flows	Shut off utilities; Evacuate building if necessary; Determine impact on the equipment and facilities and any disruption
	Lightning strike	An electrical discharge caused by lightning, typically during thunderstorms	Power off all equipment; listen to hurricane advisories; Evacuate area, if flooding is possible; Check gas, water and electrical lines for damage; Do not use telephones, in the event of severe lightning; Assess damage
	Limnic eruption	The sudden eruption of carbon dioxide from deep lake water	Shut off utilities; Evacuate building if necessary; Determine impact on the equipment and facilities and any disruption
	Tornado	Violent rotating columns of air which descent from severe thunderstorm cloud systems	Monitor tornado advisories; Power off equipment; Shut off utilities (power and gas); Assess damage once storm passes
	Tsunami	A series of water waves caused by the displacement of a large volume of a body of water, typically an ocean or a large lake, usually caused by earthquakes, volcanic eruptions, underwater explosions, landslides, glacier calvings, meteorite impacts and other disturbances above or below water	Power off all equipment; listen to tsunami advisories; Evacuate area, if flooding is possible; Check gas, water and electrical lines for damage; Assess damage

	Volcanic eruption	The release of hot magma, volcanic ash and/or gases from a volcano	Shut off utilities; Evacuate building if necessary; Determine impact on the equipment and facilities and any disruption
Man-made	Bioterrorism	The intentional release or dissemination of biological agents as a means of coercion	Get information immediately from your Public Health officials via the news media as to the right course of action; If you think you have been exposed, quickly remove your clothing and wash off your skin; Also put on a HEPA to help prevent inhalation of the agent
	Civil unrest	A disturbance caused by a group of people that may include sit-ins and other forms of obstructions, riots, sabotage and other forms of crime, and which is intended to be a demonstration to the public and the government, but can escalate into general chaos	Contact local police or law enforcement
	Fire (urban)	Even with strict building fire codes, people still perish needlessly in fires	Attempt to suppress fire in early stages; Evacuate personnel on alarm, as necessary; Notify fire department; Shut off utilities; Monitor weather advisories
	Hazardous material spills	The escape of solids, liquids, or gases that can harm people, other living organisms, property or the environment, from their intended controlled environment such as a container.	Leave the area and call the local fire department for help. If anyone was affected by the spill, call the your local Emergency Medical Services line
	Nuclear and Radiation Accidents	An event involving significant release of radioactivity to the environment or a reactor core meltdown and which leads to major undesirable consequences to people, the environment, or the facility	Recognize that a CBRN incident has or may occur. Gather, assess and disseminate all available information to first responders. Establish an overview of the affected area. Provide and obtain regular updates to and from first responders.
	Power Failure	Caused by summer or winter storms, lightning or construction equipment digging in the wrong location	Wait 5–10 minutes; Power-off all Servers after a graceful shutdown; Do not use telephones, in the event of severe lightning; Shut down main electric circuit usually located in the basement or the first floor

In the realm of information technology per se, disasters may also be the result of a computer security exploit. Some of these are: computer viruses, cyberattacks, denial-of-service attacks, hacking, and malware exploits. These are ordinarily attended to by information security experts.

Planning Methodology

According to Geoffrey H. Wold of the Disaster Recovery Journal, the entire process involved in developing a Disaster Recovery Plan consists of 10 steps:

Obtaining Top Management Commitment

For a disaster recovery plan to be successful, the central responsibility for the plan must reside on top management. Management is responsible for coordinating the disaster recovery plan and ensuring its effectiveness within the organization. It is also responsible for allocating adequate time and resources required in the development of an effective plan. Resources that management must allocate include both financial considerations and the effort of all personnel involved.

Establishing a Planning Committee

A planning committee is appointed to oversee the development and implementation of the plan. The planning committee includes representatives from all functional areas of the organization. Key committee members customarily include the operations manager and the data processing manager. The committee also defines the scope of the plan.

Performing a Risk Assessment

The planning committee prepares a risk analysis and a business impact analysis (BIA) that includes a range of possible disasters, including natural, technical and human threats. Each functional area of the organization is analyzed to determine the potential consequence and impact associated with several disaster scenarios. The risk assessment process also evaluates the safety of critical documents and vital records. Traditionally, fire has posed the greatest threat to an organization. Intentional human destruction, however, should also be considered. A thorough plan provides for the "worst case" situation: destruction of the main building. It is important to assess the impacts and consequences resulting from loss of information and services. The planning committee also analyzes the costs related to minimizing the potential exposures.

Establishing Priorities for Processing and Operations

At this point, the critical needs of each department within the organization are evaluated in order to prioritize them. Establishing priorities is important because no organization possesses infinite resources and criteria must be set as to where to allocate resources first. Some of the areas often reviewed during the prioritization process are functional operations, key personnel and their functions, information flow, processing systems used, services provided, existing documentation, historical records, and the department's policies and procedures.

Processing and operations are analyzed to determine the maximum amount of time that the department and organization can operate without each critical system. This will later get mapped into the Recovery Time Objective. A critical system is defined as that which is part of a system or procedure necessary to continue operations should a department, computer center, main facility or a combination of these be destroyed or become inaccessible. A method used to determine the critical needs of a department is to document all the functions performed by each department. Once the primary functions have been identified, the operations and processes are then ranked in order of priority: essential, important and non-essential.

Determining Recovery Strategies

During this phase, the most practical alternatives for processing in case of a disaster are researched and evaluated. All aspects of the organization are considered, including physical facilities, computer hardware and software, communications links, data files and databases, customer services provided, user operations, the overall management information systems (MIS) structure, end-user systems, and any other processing operations.

Alternatives, dependent upon the evaluation of the computer function, may include: hot sites, warm sites, cold sites, reciprocal agreements, the provision of more than one data center, the installation and deployment of multiple computer system, duplication of service center, consortium arrangements, lease of equipment, and any combinations of the above.

Written agreements for the specific recovery alternatives selected are prepared, specifying contract duration, termination conditions, system testing, cost, any special security procedures, procedure for the notification of system changes, hours of operation, the specific hardware and other equipment required for processing, personnel requirements, definition of the circumstances constituting an emergency, process to negotiate service extensions, guarantee of compatibility, availability, non-mainframe resource requirements, priorities, and other contractual issues.

Collecting Data

In this phase, data collection takes place. Among the recommended data gathering materials and documentation often included are various lists (employee backup position listing, critical telephone numbers list, master call list, master vendor list, notification checklist), inventories (communications equipment, documentation, office equipment, forms, insurance policies, workgroup and data center computer hardware, microcomputer hardware and software, office supply, off-site storage location equipment, telephones, etc.), distribution register, software and data files backup/retention schedules, temporary location specifications, any other such other lists, materials, inventories and documentation. Pre-formatted forms are often used to facilitate the data gathering process.

Organizing and Documenting a Written Plan

Next, an outline of the plan's contents is prepared to guide the development of the detailed procedures. Top management reviews and approves the proposed plan. The outline can ultimately be used for the table of contents after final revision. Other four benefits of this approach are that (1) it helps to organize the detailed procedures, (2) identifies all major steps before the actual writing process begins, (3) identifies redundant procedures that only need to be written once, and (4) provides a road map for developing the procedures.

It is often considered best practice to develop a standard format for the disaster recovery plan so as to facilitate the writing of detailed procedures and the documentation of other information to be included in the plan later. This helps ensure that the disaster plan follows a consistent format and allows for its ongoing future maintenance. Standardization is also important if more than one person is involved in writing the procedures.

It is during this phase that the actual written plan is developed in its entirety, including all detailed procedures to be used before, during, and after a disaster. The procedures include methods for maintaining and updating the plan to reflect any significant internal, external or systems changes. The procedures allow for a regular review of the plan by key personnel within the organization. The disaster recovery plan is structured using a team approach. Specific responsibilities are assigned to the appropriate team for each functional area of the organization. Teams responsible for administrative functions, facilities, logistics, user support, computer backup, restoration and other important areas in the organization are identified.

The structure of the contingency organization may not be the same as the existing organization chart. The contingency organization is usually structured with teams responsible for major functional areas such as administrative functions, facilities, logistics, user support, computer backup, restoration, and any other important area.

The management team is especially important because it coordinates the recovery process. The team assesses the disaster, activates the recovery plan, and contacts team managers. The management team also oversees, documents and monitors the recovery process. It is helpful when management team members are the final decision-makers in setting priorities, policies and procedures. Each team has specific responsibilities that are completed to ensure successful execution of the plan. The teams have an assigned manager and an alternate in case the team manager is not available. Other team members may also have specific assignments where possible.

Developing Testing Criteria and Procedures

Best practices dictate that DR plans be thoroughly tested and evaluated on a regular basis (at least annually). Thorough DR plans include documentation with the procedures for testing the plan. The tests will provide the organization with the assurance that all necessary steps are included in the plan. Other reasons for testing include:

- Determining the feasibility and compatibility of backup facilities and procedures.
- Identifying areas in the plan that need modification.
- Providing training to the team managers and team members.
- Demonstrating the ability of the organization to recover.
- Providing motivation for maintaining and updating the disaster recovery plan.

Testing the Plan

After testing procedures have been completed, an initial "dry run" of the plan is performed by conducting a structured walk-through test. The test will provide additional information regarding any further steps that may need to be included, changes in procedures that are not effective, and other appropriate adjustments. These may not become evident unless an actual dry-run test is performed. The plan is subsequently updated to correct any problems identified during the test. Initially, testing of the plan is done in sections and after normal business hours to minimize disruptions to the overall operations of the organization. As the plan is further polished, future tests occur during normal business hours.

Types of tests include: checklist tests, simulation tests, parallel tests, and full interruption tests.

Obtaining Plan Approval

Once the disaster recovery plan has been written and tested, the plan is then submitted to management for approval. It is top management's ultimate responsibility that the organization has a documented and tested plan. Management is responsible for (1) establishing the policies, procedures and responsibilities for comprehensive contingency planning, and (2) reviewing and approving the contingency plan annually, documenting such reviews in writing.

Organizations that receive information processing from service bureaus will, in addition, also need to (1) evaluate the adequacy of contingency plans for its service bureau, and (2)ensure that its contingency plan is compatible with its service bureau's plan.

Measuring Plan Success

Many communities utilize online planning tools to better prepare for and track their recovery in the wake of natural disasters. The Disaster Recovery Tracking Tool is designed to assist local officials and recovery specialists in tracking progress toward recovery goals by identifying relevant metrics that can be used to monitor changes over time. The Tool's themes, focus areas, and metrics were derived from a study conducted through a joint effort by the University of North Carolina at Chapel Hill's U.S. Department of Homeland Security Coastal Hazards Center of Excellence and Texas A&M University Health Science Center School of Public Health. It allows communities to leverage existing statistics to better plan for and understand community recovery. <http://www.trackyourrecovery.org/>

Caveats/Controversies

Due to its high cost, disaster recovery plans are not without critics. Cormac Foster has identified five "common mistakes" organizations often make related to disaster recovery planning:

Lack of Buy-in

One factor is the perception by executive management that DR planning is "just another fake earthquake drill" or CEOs that fail to make DR planning and preparation a priority, are often significant contributors to the failure of a DR plan.

Incomplete RTOs and RPOs

Another critical point is failure to include each and every important business process or a block of data. "Every item in your DR plan requires a Recovery Time Objective (RTO) defining maximum process downtime or a Recovery Point Objective (RPO) noting an acceptable restore point. Anything less creates ripples that can extend the disaster's impact." As an example, "payroll, accounting and the weekly customer newsletter may not be mission-critical in the first 24 hours, but left alone for several days, they can become more important than any of your initial problems."

Systems Myopia

A third point of failure involves focusing only on DR without considering the larger business continuity needs: "Data and systems restoration after a disaster are essential, but every business process in your organization will need IT support, and that support requires planning and resources." As an example, corporate office space lost to a disaster can result in an instant pool of teleworkers which, in turn, can overload a company's VPN overnight, overwork the IT support staff at the blink of an eye and cause serious bottlenecks and monopolies with the dial-in PBX system.

Lax Security

When there is a disaster, an organization's data and business processes become vulnerable. As such, security can be more important than the raw speed involved in a disaster recovery plan's RTO. The most critical consideration then becomes securing the new data pipelines: from new VPNs to the connection from offsite backup services. Another security concern includes documenting every step of the recovery process—something that is especially important in highly regulated industries, government agencies, or in disasters requiring post-mortem forensics. Locking down or remotely wiping lost handheld devices is also an area that may require addressing.

Outdated Plans

Another important aspect that is often overlooked involves the frequency with which DR Plans are updated. Yearly updates are recommended but some industries or organizations require more frequent updates because business processes evolve or because of quicker data growth. To stay relevant, disaster recovery plans should be an integral part of all business analysis processes, and should be revisited at every major corporate acquisition, at every new product launch and at every new system development milestone.

Bug-out Bag

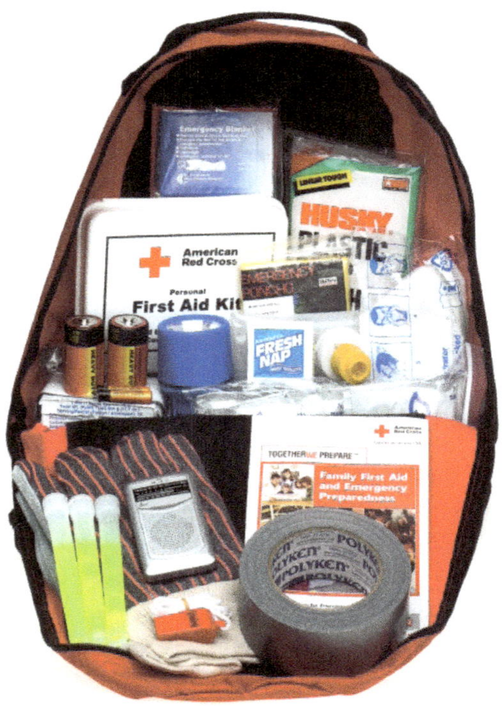

Off-the-shelf Red Cross preparedness kit .

A bug-out bag is a portable kit that normally contains the items one would require to survive for seventy-two hours when evacuating from a disaster, however some kits are designed to last longer periods of time than just 72 hours. The focus is on evacuation, rather than long-term survival, distinguishing the bug-out bag from a survival kit, a boating or aviation emergency kit, or a fixed-site disaster supplies kit. The kits are also popular in the survivalism and prepper subcultures.

The term "bug-out bag" is related to, and possibly derived from, the "bail-out bag" emergency kit many military aviators carry. In the United States, the term refers to the Korean War practice of the U.S. Army designating alternative defensive positions, in the event that the units had to displace. They were directed to "bug out" when being overrun was imminent. The concept passed into wide usage among other military and law enforcement personnel, though the "bail-out bag" is as likely to include emergency gear for going into an emergency situation as for escaping an emergency.

Other names for such a bag are a BOB, 72-Hour Kit, a Grab Bag, a Battle Box, a Personal Emergency Relocation Kits (PERK), a Go Bag, a GOOD Bag (Get Out Of Dodge) or INCH Bag (I'm Never Coming Home).

The term *Go-Kit* is popular in the amateur radio service, especially in the "Radio Amateur Civil Emergency Service" (RACES) and "Amateur Radio Emergency Service" (ARES) communities and is used to describe a combination personal bug-out-bag and portable amateur radio station. A personal Go-Kit generally takes some combination of units - a "one-day" (or "24 hour") kit, a "three day" (or "72 hour") kit that adds additional supplies, or a "one week kit" that adds yet additional personal items to the three-day kit. Any or all supports deploying the operator plus his or her privately owned self-contained radio communications setup.

Rationale

The primary purpose of a bug-out bag is to allow one to evacuate quickly if a disaster should strike. It is therefore prudent to gather all of the materials and supplies that might be required to do this into a single place, such as a bag or a few storage containers. The recommendation that a bug-out bag contain enough supplies for seventy-two hours arises from advice from organizations responsible for disaster relief and management that it may take them up to seventy-two hours to reach people affected by a disaster and offer help. The bag's contents may vary according to the region of the user, as someone evacuating from the path of a hurricane may have different supplies from someone who lives in an area prone to blizzards, tornadoes or wildfires.

In addition to allowing one to survive a disaster evacuation, a bug-out bag may also be used when sheltering in place as a response to emergencies such as house fires, blackouts, tornadoes, and other severe natural disasters.

Some survivalists also recommend keeping a 'get me home' kit in the car and/or at work. This is a kit to enable a person to get back home from work in an emergency where all transport cars and public transport have broken down. It is designed around personal circumstances where, for example, a walk of 25 kilometres might be required from work to home. The 'get me home' kit can include, for example, enough water to get home, suitable walking shoes, a map (not electronic), enough food for 12 hours, clothing for adverse weather, etc.

Typical Contents

The suggested contents of a bug-out bag vary, but most of the following are usually included:

- Enough food and water to last for at least 72 hours. This includes:
    - Water for washing, drinking and cooking. Canada recommends 2 litres per person per day for drinking plus an additional 2 litres per person per day for cleaning and hygiene. New Zealand recommends 3 litres per person per day for drinking. US recommends 1 gallon (3.78 litres) per person per day. If you have pets then include extra water for them.
    - Non-perishable food
    - Water purification supplies.
    - Cooking supplies.
- A first aid kit.
- Fire starting tool (e.g., matches, ferrocerium rod, lighter, 9volt battery, etc.).
- A disaster plan including location of emergency centers, rallying points, possible evacuation routes, etc.
- Professional emergency literature explaining what to do in various types of disaster, studied and understood before the actual disaster but kept for reference.
- Maps and travel information.
- Standard camping equipment, including sanitation supplies.
- Weather appropriate clothing and UV protection (e.g., poncho, golf umbrella headwear, gloves, etc.).
- Bedding items such as sleeping bags and blankets. If you have to sleep on the ground then a thick plastic or oilcloth tarp under the sleeping bags will keep ground moisture out of your bedding. A second one suspended on a rope between two trees or other supports can be used to form a tarp tent shelter from the morning dew or a rainstorm.
- Enough medicine to last an extended evacuation period.
- Copies of medical records for each person in the family.
- Pet, child, and elderly care needs.
- Battery or crank-operated radio.
- Lighting (battery or crank operated flashlight, glow sticks).
- Cash and change, as electronic banking transactions may not be available during the initial period following an emergency or evacuation.

- Positive identification, such as drivers license, state I.D. card, or social security card, plus any medical ID cards if you have them. Those with allergies should have a MedicAlert or similar ID.
- Birth certificate or passport.
- Fixed-blade and folding knife
- Swiss Army Knife
- Multi-tool, like a Leatherman
- Duct tape and rope or paracord
- Tarpaulins for shelter and water collection
- Wire for binding and animal traps
- Compass
- Firearms and extra ammunition
- Slingshot, pellet gun, blowgun or other small game hunting equipment
- Small fishing kit
- Signal mirror
- Emergency whistle
- Rubber tubing
- Digestion care medicine for indigestion, stomach ache, nausea, and diarrhea

Dam Safety System

Dam safety systems are systems monitoring the state of dams used for hydropower or other purposes, as well as external physical threats to them, and issuing emergency warnings at various degrees of automation. This includes the use of differential GPS and SAR remote sensing to monitor the risks imposed by landslides and subsidence. For large dams seismographs are used to detect reservoir-induced seismicity that could threaten the stability of the dam. The output of these systems can provide warning to the local population ahead of a potential collapse.

Particular Applications

Ukraine

The monitoring and warning system safeguarding the Kiev Reservoir dam is complicated enough to include protection from a space object impact.

United States

Dam safety systems became a focus of multi-agency regulations during the U.S. Army Corps of Engineers construction of large flood control and hydro-electric power generation projects. To help benchmark proven practices, the Association of State Dam Safety Officials (ASDSO) formed a national non-profit organization of state and federal dam safety regulators, dam owners and operators, engineering consultants, manufacturers and suppliers, academia, contractors and others interested in dams safety. More recently public safety concerns were addressed by the Indian Dams Safety Act of 1992 during hearings before the Select Committee on Indian Affairs, United States Senate, 102nd Congress, second session, on S. 2617. The purpose was to provide for the maintenance of dams located on Indian lands in New Mexico by the Bureau of Indian Affairs through contracts with the Indian tribes. (August 4, 1992 in Washington, D.C.)

The ASDSO Conference Proceedings paper by Gary R. Holtzhausen (1991) describes the effective use of tiltmeters with remote sensing to provide reliable low-cost early warning of impending structural failures.

The ASDSO Conference Proceedings paper by Barry K. Meyers (2002) describes two case studies using failure modes analysis together with a variety of automated instrumentation to provide early warnings at White River Project owned by Puget Sound Energy as well as a case study of the Silver Creek Dam near Silverton OR.

Tsunami Warning System

Evacuation route sign in a low-lying coastal area on the West Coast of the United States

A tsunami warning system (TWS) is used to detect tsunamis in advance and issue warnings to prevent loss of life and damage. It is made up of two equally important components: a network of sensors to detect tsunamis and a communications infrastructure to issue timely alarms to permit evacuation of the coastal areas. There are two distinct types of tsunami warning systems: international and regional. When operating, seismic alerts are used to instigate the watches and warnings; then, data from observed sea level height (either shore-based tide gauges or DART buoys) are used to verify the existence of a tsunami. Other systems have been proposed to augment the warning procedures; for example, it has been suggested that the duration and frequency content of t-wave energy (which is earthquake energy trapped in the ocean SOFAR channel) is indicative of an earthquake's tsunami potential.

History and Forecasting

The first rudimentary system to alert communities of an impending tsunami was attempted in Hawaii in the 1920s. More advanced systems were developed in the wake of the April 1, 1946 (caused by the 1946 Aleutian Islands earthquake) and May 23, 1960 (caused by the 1960 Valdivia earthquake) tsunamis which caused massive devastation in Hilo, Hawaii. While tsunamis travel at between 500 and 1,000 km/h (around 0.14 and 0.28 km/s) in open water, earthquakes can be detected almost at once as seismic waves travel with a typical speed of 4 km/s (around 14,400 km/h). This gives time for a possible tsunami forecast to be made and warnings to be issued to threatened areas, if warranted. Unfortunately, until a reliable model is able to predict which earthquakes will produce significant tsunamis, this approach will produce many more false alarms than verified warnings.

International Warning Systems (IWS)

Pacific Ocean

Japanese Tsunami warning sign

Tsunami warnings for most of the Pacific Ocean are issued by the Pacific Tsunami Warning Center (PTWC), operated by the United States NOAA in Ewa Beach, Hawaii. NOAA's National Tsunami Warning Center (NTWC) in Palmer, Alaska issues warnings for the west coast of North America, including Alaska, British Columbia, Oregon and California. The PTWC was established in 1949, following the 1946 Aleutian Island earthquake and a tsunami that resulted in 165 casualties on Hawaii and in Alaska; NTWC was founded in 1967. International coordination is achieved through the International Coordination Group for the Tsunami Warning System in the Pacific, established by the Intergovernmental Oceanographic Commission of UNESCO.

Indian Ocean (ICG/IOTWS)

After the 2004 Indian Ocean Tsunami which killed almost 250,000 people, a United Nations conference was held in January 2005 in Kobe, Japan, and decided that as an initial step towards an International Early Warning Programme, the UN should establish an Indian Ocean Tsunami Warning System. This then resulted in a system of warnings in Indonesia and other affected areas.

Tsunami Early Warning Tower board in Hikkaduwa, Sri Lanka

North Eastern Atlantic, the Mediterranean and Connected Seas (ICG/NEAMTWS)

The First United Session of the Inter-governmental Coordination Group for the Tsunami Early Warning and Mitigation System in the North Eastern Atlantic, the Mediterranean and connected Seas (ICG/NEAMTWS), established by the Intergovernmental Oceanographic Commission of UNESCO Assembly during its 23rd Session in June 2005, through Resolution XXIII.14, took place in Rome on 21 and 22 November 2005.

The meeting, hosted by the Government of Italy (the Italian Ministry of Foreign Affairs and the Italian Ministry for the Environment and Protection of Land and Sea), was attended by more than 150 participants from 24 countries, 13 organizations and numerous observers.

Caribbean

A Caribbean-wide tsunami warning system was planned to be instituted by the year 2010, by representatives of Caribbean nations who met in Panama City in March 2008. Panama's last major tsunami killed 4,500 people in 1882. Barbados has said it will review or test its tsunami protocol in February 2010 as a regional pilot.[needs update]

Regional Warning Systems

Regional (or local) warning system centres use seismic data about nearby recent earthquakes to determine if there is a possible local threat of a tsunami. Such systems are capable of issuing warnings to the general public (via public address systems and sirens) in less than 15 minutes. Although the epicenter and moment magnitude of an underwater quake and the probable tsunami arrival times can be quickly calculated, it is almost always impossible to know whether underwater ground shifts have occurred which will result in tsunami waves. As a result, false alarms can occur with these systems, but the disruption is small, which makes sense due to the highly localised nature of these extremely quick warnings, in combination with how difficult it would be for a false alarm to affect more than a small area of the system. Real tsunamis would affect more than just a small portion.

Japan

Japan has a nationwide tsunami warning system. The system usually issues the warning minutes after an Earthquake Early Warning (EEW) is issued, should there be expected waves. The tsunami warning was issued within 3 minutes with the most serious rating on its warning scale during the 2011 Tōhoku earthquake and tsunami; it was rated as a "major tsunami", being at least 3 m (9.8 ft) high. An improved system was unveiled on March 7, 2013 following the 2011 disaster to better access the waves.

Conveying the Warning

Mast with warning system, and sign detailing escape routes, on the coast of Okumatsushima, Miyagi prefecture, Japan (this coast was severely hit by the 2011 tsunami)

Detection and prediction of tsunamis is only half the work of the system. Of equal importance is the ability to warn the populations of the areas that will be affected. All tsunami warning systems feature multiple lines of communications (such as SMS, e-mail, fax, radio, texting and telex, often using hardened dedicated systems) enabling emergency messages to be sent to the emergency services and armed forces, as well to population-alerting systems (e.g. sirens) and the Emergency Alert System

Shortcomings

With the speed at which tsunami waves travel through open water, no system can protect against a very sudden tsunami, where the coast in question is too close to the epicenter. A devastating tsunami occurred off the coast of Hokkaidō in Japan as a result of an earthquake on July 12, 1993. As a result, 202 people on the small island of Okushiri, Hokkaido lost their lives, and hundreds more were missing or injured. This tsunami struck just three to five minutes after the quake, and most victims were caught while fleeing for higher ground and secure places after surviving the earthquake. This was also the case in Aceh, Indonesia.

While there remains the potential for sudden devastation from a tsunami, warning systems can be effective. For example, if there were a very large subduction zone earthquake (moment magnitude 9.0) off the west coast of the United States, people in Japan, would therefore have more than 12 hours (and likely warnings from warning systems in Hawaii and elsewhere) before any tsunami arrived, giving them some time to evacuate areas likely to be affected.

Indian Ocean Tsunami Warning System

The Indian Ocean Tsunami Warning System is a tsunami warning system set up to provide warning to inhabitants of nations bordering the Indian Ocean of approaching tsunamis.

Background

A warning system for the Indian Ocean was prompted by the 2004 Indian Ocean earthquake and resulting tsunami, which left some 220,000 people dead or missing. Many analysts claimed that the disaster would have been mitigated if there had been an effective warning system in place, citing the well-established Hawaii-based Pacific Tsunami Warning Center, which operates in the Pacific Ocean.

People in some areas would have had more than adequate time to seek safety if they were aware of the impending catastrophe. The only way to effectively mitigate the impact of a tsunami is through an early warning system. Other methods such as sea walls only work for a percentage of waves, but a warning system is effective for all waves originating outside a minimum distance from the coastline.

Construction

The Indian Ocean Tsunami Warning System was agreed to in a United Nations conference held in January 2005 in Kobe, Japan as an initial step towards an International Early Warning Programme. Nanometrics (Ottawa, Canada) and RESULTS Marine Private Limited, India, delivered and successfully installed 17 Seismic VSAT stations with 2 Central Recording Station to provide the seismic event alert to the scientists through SMS and E-mail automatically within 2 min.

The system became active in late June 2006 following the leadership of UNESCO. It consists of 25 seismographic stations relaying information to 26 national tsunami information centers, as well as 6 Deep-ocean Assessment and Reporting of Tsunami (DART) buoys. However, UNESCO warned that further coordination between governments and methods of relaying information from the centers to the civilians at risk are required to make the system effective.

Sensor data is processed by the U.S. Pacific Tsunami Warning Center in Hawaii and the Japan Meteorological Agency, and alerts are forwarded to threatened countries and also made available to the general public. National governments warn citizens through a variety of means, including SMS messages, radio and television broadcasts, sirens from dedicated platforms and mosque loudspeakers, and police vehicles with loudspeakers.

Performance During Emergencies

The system was not yet operational during the 2006 Pangandaran earthquake and tsunami. The Indonesian government did receive tsunami warnings from the warning centers but did not have a system to relay the alert to its citizens. At least 23,000 people did evacuate the coast after the quake, either fearing a tsunami or because their homes had been destroyed. Waves as high as 7.39 m (24.2 ft) still resulted in about 700 fatalities and 9,000 injuries.

In the 2012 Banda Aceh earthquake of 8.4 magnitude, the system alerted the Indian islands on Andaman and Nicobar within 8 minutes. Some tsunami warning sirens in Aceh were delayed by about 20 minutes due to failure of the electrical grid caused by the proximity of the earthquake, and evacuation routes in Banda Aceh were jammed with traffic.

Of the 28 countries that ring the Indian Ocean, now Australia, Indonesia and India are responsible for spearheading tsunami warnings in the area.

Community Resilience

Community resilience is the sustained ability of a community to utilize available resources (e.g. energy, communication, food, etc.) to respond to, withstand, and recover from adverse situations (e.g. economic collapse to global catastrophic risks. This allows for the evolution and growth of a community after disaster strikes. Communities that are resilient are able to minimize any disaster, making the return to normal life as effortless as possible. By implementing a community resilience plan, a community can come together and overcome any disaster, while rebuilding physically and economically.

What is a Community Resilience Plan?

A community resilience plan is a plan of action that allows for a community to rebuild after disaster. Each plan should entail specific guidelines that will aid the community to rebuild both the economy and the ecosystem that the community thrives on. this typically means there are measures in place that a community will follow, such as the distribution of volunteers, and the access to knowledge necessary to rebuild.

References

- ICC/NSSA Standard for the Design and Construction of Storm Shelters (First ed.). International Code Council, Inc. August 2008. ISBN 978-1-58001-746-6.
- Kearny, Cresson H (1986). Nuclear War Survival Skills. Oak Ridge, TN: Oak Ridge National Laboratory. pp. 6–10. ISBN 0-942487-01-X.
- Springer, Chris (2003). Pyongyang: the hidden history of the North Korean capital. Entente Bt. p. 125. ISBN 978-963-00-8104

- Reitherman, Robert (2012). Earthquakes and Engineers: An International History. Reston, VA: ASCE Press. pp. 486–487. ISBN 9780784410714.
- South, J. Allan (1990). “Chapter 11 (Equipment), Bug-Out Bag Contents”. The Sense of Survival. Orem, Utah: Timpanogos Publishers. p. 221. ISBN 0-935329-00-5.
- Stewart, Creek (2012). Build the Perfect Bug Out Bag: Your 72-Hour Disaster Survival Kit. Betterway Books. ISBN 1440318743.

5

Emergency Management: A Global Overview

To develop a global understanding of emergency, it is important to mention agencies such as United States Environmental Protection Agency, Local Emergency Planning Committee, the World Conference on Disaster Risk Reduction and Federal Disaster Assistance Nonprofit Fairness Act of 2013. This chapter is an overview of the subject matter incorporating all the major aspects of emergency management.

United States Environmental Protection Agency

The United States Environmental Protection Agency (EPA or sometimes USEPA) is an agency of the Federal government of the United States which was created for the purpose of protecting human health and the environment by writing and enforcing regulations based on laws passed by Congress. The EPA was proposed by President Richard Nixon and began operation on December 2, 1970, after Nixon signed an executive order. The order establishing the EPA was ratified by committee hearings in the House and Senate. The agency is led by its Administrator, who is appointed by the president and approved by Congress. The current administrator is Gina McCarthy. The EPA is not a Cabinet department, but the administrator is normally given cabinet rank.

The EPA has its headquarters in Washington, D.C., regional offices for each of the agency's ten regions, and 27 laboratories. The agency conducts environmental assessment, research, and education. It has the responsibility of maintaining and enforcing national standards under a variety of environmental laws, in consultation with state, tribal, and local governments. It delegates some permitting, monitoring, and enforcement responsibility to U.S. states and the federally recognized tribes. EPA enforcement powers include fines, sanctions, and other measures. The agency also works with industries and all levels of government in a wide variety of voluntary pollution prevention programs and energy conservation efforts.

The agency has approximately 15,193 full-time employees and engages many more people on a contractual basis. More than half of EPA human resources are engineers, scientists, and environmental protection specialists; other groups include legal, public affairs, financial, and information technologists.

History

Stacks emitting smoke from burning discarded automobile batteries, photo taken in Houston in 1972 by Marc St. Gil (cs), official photographer of recently founded EPA

Same smokestacks in 1975 after the plant was closed in a push for greater environmental protection

Beginning in the late 1950s and through the 1960s, Congress reacted to increasing public concern about the impact that human activity could have on the environment. A key legislative option to address this concern was the declaration of a national environmental policy. Advocates of this approach argued that without a specific policy, federal agencies were neither able nor inclined to consider the environmental impacts of their actions in fulfilling the agency's mission. The statute that ultimately addressed this issue was the National Environmental Policy Act of 1969 (NEPA, 42 U.S.C. §§ 4321–4347). Senator Henry M. Jackson proposed and helped write S 1075, the bill that eventually became the National Environmental Policy Act. The law was signed by President Nixon on January 1, 1970. NEPA was the first of several major environmental laws passed in the 1970s. It declared a national policy to protect the environment and created a Council on Environmental Quality (CEQ) in the Executive Office of the President. To implement the national policy, NEPA required that a detailed statement of environmental impacts be prepared for all major federal actions significantly affecting the environment. The "detailed statement" would ultimately be referred to as an environmental impact statement (EIS).

Ruckelshaus sworn in as first EPA Administrator.

In 1970, President Richard Nixon proposed an executive reorganization that would consolidate many of the federal government's environmental responsibilities under one agency, a new Environmental Protection Agency. That reorganization proposal was reviewed and passed by the House and Senate.

Staff recall that in the early days there was "an enormous sense of purpose and excitement" and the expectation that "there was this agency which was going to do something about a problem that clearly was on the minds of a lot of people in this country," leading to tens of thousands of resumes from those eager to participate in the mighty effort to clean up America's environment.

For at least 10 years before NEPA was enacted, Congress debated issues that the act would ultimately address. The act was modeled on the Resources and Conservation Act of 1959, introduced by Senator James E. Murray in the 86th Congress. That bill would have established an environmental advisory counsel in the office of the President, declared a national environmental policy, and required the preparation of an annual environmental report. In the years following the introduction of Senator Murray's bill, similar bills were introduced and hearings were held to discuss the state of the environment and Congress's potential responses to perceived problems. In 1968, a joint House-Senate colloquium was convened by the chairmen of the Senate Committee on Interior and Insular Affairs (Senator Henry M. Jackson) and the House Committee on Science and Astronautics (Representative George Miller) to discuss the need for and potential means of implementing a national environmental policy. In the colloquium, some Members of Congress expressed a continuing concern over federal agency actions affecting the environment.

When EPA first began, the private sector felt strongly that the environmental protection movement was a passing fad, and the agency's first administrator, William Ruckelshaus, felt pressure to show a public that was deeply skeptical about government's effectiveness that EPA could respond effectively to widespread concerns about pollution.

The EPA began regulating greenhouse gases (GHGs) from mobile and stationary sources of air pollution under the Clean Air Act (CAA) for the first time on January 2, 2011. Standards for mobile sources have been established pursuant to Section 202 of the CAA, and GHGs from stationary sources are controlled under the authority of Part C of Title I of the Act.

In May 2013, Congress renamed the EPA headquarters as the William Jefferson Clinton Federal Building, after former president Bill Clinton.

During the 2016 Presidental Election Republican Canidates, Ted Cruz, Marco Rubio and Donald Trump critiziced the EPA of lossing jobs, taking states rights to regulate green house gases away.

Organization

Offices

- Office of the Administrator (OA)
- Office of Administration and Resources Management (OARM)
- Office of Air and Radiation (OAR)
- Office of Chemical Safety and Pollution Prevention (OCSPP)
- Office of the Chief Financial Officer (OCFO)
- Office of Enforcement and Compliance Assurance (OECA)
- Office of Environmental Information (OEI)
- Office of General Counsel (OGC)
- Office of Inspector General (OIG)
- Office of International and Tribal Affairs (OITA)
- Office of Research and Development (ORD)
- Office of Land and Emergency Management (OLEM)
- Office of Water (OW)

Regions

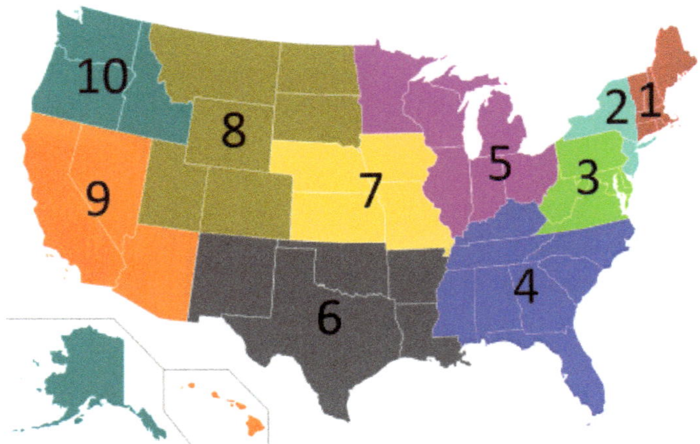

The administrative regions of the United States Environmental Protection Agency.

Creating 10 EPA regions was an initiative that came from President Richard Nixon.

Each EPA regional office is responsible within its states for implementing the Agency's programs, except those programs that have been specifically delegated to states.

- Region 1: responsible within the states of Connecticut, Maine, Massachusetts, New Hampshire, Rhode Island, and Vermont (New England).
- Region 2: responsible within the states of New Jersey and New York. It is also responsible for the US territories of Puerto Rico, and the U.S. Virgin Islands.
- Region 3: responsible within the states of Delaware, Maryland, Pennsylvania, Virginia, West Virginia, and the District of Columbia.
- Region 4: responsible within the states of Alabama, Florida, Georgia, Kentucky, Mississippi, North Carolina, South Carolina, and Tennessee.
- Region 5: responsible within the states of Illinois, Indiana, Michigan, Minnesota, Ohio, and Wisconsin.
- Region 6: responsible within the states of Arkansas, Louisiana, New Mexico, Oklahoma, and Texas.
- Region 7: responsible within the states of Iowa, Kansas, Missouri, and Nebraska.
- Region 8: responsible within the states of Colorado, Montana, North Dakota, South Dakota, Utah, and Wyoming.
- Region 9: responsible within the states of Arizona, California, Hawai'i, Nevada, the territories of Guam and American Samoa, and the Navajo Nation.
- Region 10: responsible within the states of Alaska, Idaho, Oregon, and Washington.

Each regional office also implements programs on Indian Tribal lands, except those programs delegated to Tribal authorities.

Related Legislation

The legislation here is general environmental protection legislation, and may also apply to other units of the government, including the Department of the Interior and the Department of Agriculture.

Air

- 1955: Air Pollution Control Act PL 84-159
- 1963: Clean Air Act PL 88-206
- 1965: Motor Vehicle Air Pollution Control Act PL 89-272
- 1966: Clean Air Act Amendments PL 89-675
- 1967: Air Quality Act PL 90-148
- 1970: Clean Air Act Extension PL 91-604
- 1977: Clean Air Act Amendments PL 95-95
- 1990: Clean Air Act Amendments PL 101-549

Water

- 1948: Water Pollution Control Act PL 80-845
- 1965: Water Quality Act PL 89-234
- 1966: Clean Waters Restoration Act PL 89-753
- 1970: Water Quality Improvement Act PL 91-224
- 1972: Federal Water Pollution Control Amendments of 1972 PL 92-500
- 1974: Safe Drinking Water Act PL 93-523
- 1977: Clean Water Act PL 95-217
- 1987: Water Quality Act PL 100-4
- 1996: Safe Drinking Water Act Amendments of 1996

Land

- 1964: Wilderness Act PL 88-577
- 1968: Wild and Scenic Rivers Act PL 90-542
- 1970: Wilderness Act PL 91-504
- 1977: Surface Mining Control and Reclamation Act PL 95-87

- 1978: Wilderness Act PL 98-625
- 1980: Alaska National Interest Lands Conservation Act PL 96-487
- 1994: California Desert Protection Act PL 103-433
- 2010: California Desert Protection Act

Endangered Species

- 1946: Fish and Wildlife Coordination Act PL 79-732
- 1966: Endangered Species Preservation Act PL 89-669
- 1969: Endangered Species Conservation Act PL 91-135
- 1972: Marine Mammal Protection Act PL 92-522
- 1973: Endangered Species Act PL 93-205
- 1979: Endangered Species Preservation Act PL 95 335

Hazardous Waste

- 1965: Solid Waste Disposal Act PL 89-272
- 1970: Resource Recovery Act PL 91-512
- 1976: Resource Conservation and Recovery Act PL 94-580
- 1980: Comprehensive Environmental Response, Compensation, and Liability Act ("Superfund") PL 96-510
- 1984: Hazardous and Solid Wastes Amendments Act PL 98-616
- 1986: Superfund Amendments and Reauthorization Act PL 99-499
- 2002: Small Business Liability Relief and Brownfields Revitalization Act ("Brownfields Law") PL 107-118

Other

- 1947: Federal Insecticide, Fungicide, and Rodenticide Act PL 80-104
- 1969: National Environmental Policy Act PL 91-190
- 1972: Federal Environmental Pesticide Control Act PL 92-516
- 1976: Toxic Substances Control Act PL 94-469
- 1982: Nuclear Waste Repository Act PL 97-425
- 1996: Food Quality Protection Act PL 104-170

Programs

A bulldozer piles boulders in an attempt to prevent lake shore erosion, 1973 (photograph by Paul Sequeira, photojournalist and contributing photographer to the Environmental Protection Agency's DOCUMERICA project in the early 1970s)

It is worth noting that, in looking back in 2013 on the agency he helped shape from the beginning, Administrator William Ruckelshaus observed that a danger for EPA was that air, water, waste and other programs would be unconnected, placed in "silos," a problem that persists more than 50 years later, albeit less so than at the start.

EPA Safer Choice

The EPA Safer Choice label, previously known as the Design for the Environment (DfE) label, helps consumers and commercial buyers identify and select products with safer chemical ingredients, without sacrificing quality or performance. When a product has the Safer Choice label, it means that every intentionally - added ingredient in the product has been evaluated by EPA scientists. Only the safest possible functional ingredients are allowed in products with the Safer Choice label.

Safer Detergents Stewardship Initiative

Through the Safer Detergents Stewardship Initiative (SDSI), EPA's Design for the Environment (DfE) recognizes environmental leaders who voluntarily commit to the use of safer surfactants. Safer surfactants are the ones that break down quickly to non-polluting compounds and help protect aquatic life in both fresh and salt water. Nonylphenol ethoxylates, commonly referred to as NPEs, are an example of a surfactant class that does not meet the definition of a safer surfactant.

The Design for the Environment, which was renamed to EPA Safer Choice in 2015, has identified safer alternative surfactants through partnerships with industry and environmental advocates. These safer alternatives are comparable in cost and are readily available. CleanGredients is a source of safer surfactants.

Energy Star

In 1992 the EPA launched the Energy Star program, a voluntary program that fosters energy efficiency.

SmartGrowth

EPA's Smart Growth Program helps communities improve their development practices and get the type of development they want. EPA works with local, state, and national experts to discover and encourage development strategies that protect human health and the environment, create economic opportunities, and provide attractive and affordable neighborhoods for people of all income levels.

Pesticides

EPA administers the Federal Insecticide, Fungicide, and Rodenticide Act (FIFRA) (which is much older than the agency) and registers all pesticides legally sold in the United States.

Fuel Economy

Manufacturers selling automobiles in the United States are required to provide EPA fuel economy test results for their vehicles and the manufacturers are not allowed to provide results from alternate sources. The fuel economy is calculated using the emissions data collected during two of the vehicle's Clean Air Act certification tests by measuring the total volume of carbon captured from the exhaust during the tests.

The current testing system was originally developed in 1972 and used driving cycles designed to simulate driving during rush-hour in Los Angeles during that era. Prior to 1984 the EPA reported the exact fuel economy figures calculated from the test. In 1984, the EPA began adjusting city (aka Urban Dynamometer Driving Schedule or UDDS) results downward by 10% and highway (aka HighWay Fuel Economy Test or HWFET) results by 22% to compensate for changes in driving conditions since 1972 and to better correlate the EPA test results with real-world driving. In 1996, the EPA proposed updating the Federal Testing Procedures to add a new higher speed test (US06) and an air-conditioner on test (SC03) to further improve the correlation of fuel economy and emission estimates with real-world reports. The updated testing methodology was finalized in December, 2006 for implementation with model year 2008 vehicles and set the precedent of a 12-year review cycle for the test procedures.

In February 2005, the organization launched a program called "Your MPG" that allows drivers to add real-world fuel economy statistics into a database on the EPA's fuel economy website and compare them with others and the original EPA test results.

It is important to note that the EPA actually conducts these tests on very few vehicles. "While the public mistakenly presumes that this federal agency is hard at work conducting complicated tests on every new model of truck, van, car, and SUV, in reality, just 18 of the EPA's 17,000 employees work in the automobile-testing department in Ann Arbor, Michigan, examining 200 to 250 vehicles a year, or roughly 15 percent of new models. As to that other 85 percent, the EPA takes automakers at their word—without any testing-accepting submitted results as accurate." Two-thirds of the vehicles the EPA tests themselves are selected randomly, and the remaining third are tested for specific reasons.

Although originally created as a reference point for fossil fuelled vehicles, driving cycles have been used for estimating how many miles an electric vehicle will do on a single charge.

Air Quality

The Air Quality Modeling Group (AQMG) is in the EPA's Office of Air and Radiation (OAR) and provides leadership and direction on the full range of air quality models, air pollution dispersion models and other mathematical simulation techniques used in assessing pollution control strategies and the impacts of air pollution sources.

The AQMG serves as the focal point on air pollution modeling techniques for other EPA headquarters staff, EPA regional Offices, and State and local environmental agencies. It coordinates with the EPA's Office of Research and Development (ORD) on the development of new models and techniques, as well as wider issues of atmospheric research. Finally, the AQMG conducts modeling analyses to support the policy and regulatory decisions of the EPA's Office of Air Quality Planning and Standards (OAQPS).

The AQMG is located in Research Triangle Park, North Carolina.

Addressing air quality issues helps diminish the risk of pollution-related diseases.

Oil Pollution

The Spill Prevention, Control, and Countermeasure Rule applies to all facilities that store, handle, process, gather, transfer, store, refine, distribute, use or consume oil or oil products. Oil products includes petroleum and non-petroleum oils as well as: animal fats, oils and greases; fish and marine mammal oils; and vegetable oils, (including oils from seeds, nuts, fruits, and kernels). Mandates that a written plan is required for facilities that store more than 1,320 gallons of fuel above ground or more than 42,000 gallons below-ground, and may reasonably be expected to discharge to navigable waters(as defined in the Clean Water Act)or adjoining shorelines. Secondary Containment mandated at oil storage facilities. Oil release containment is required at oil development sites.

Toxics Release Inventory

The Toxics Release Inventory (TRI) is a resource for learning about toxic chemical releases and pollution prevention activities reported by industrial and federal facilities. TRI data support informed decision-making by communities, government agencies, companies, and others.

EPA WaterSense

WaterSense is an EPA program designed to encourage water efficiency in the United States through the use of a special label on consumer products. It was launched in June 2006. Products include high-efficiency toilets (HETs), bathroom sink faucets (and accessories), and irrigation equipment. WaterSense is a voluntary program, with EPA developing specifications for water-efficient products through a public process and product testing by independent laboratories.

Drinking Water

EPA ensures safe drinking water for the public, by setting standards for more than 160,000 public water systems nationwide. EPA oversees states, local governments and water suppliers to enforce the standards under the Safe Drinking Water Act. The program includes regulation of injection wells in order to protect underground sources of drinking water. Select readings of amounts of certain contaminants in drinking water, precipitation, and surface water, in addition to milk and air, are reported on EPA's Rad Net web site in a section entitled Envirofacts. In certain cases, readings exceeding EPA MCL levels are deleted or not included despite mandatory reporting regulations. A draft of revised EPA regulations relaxes the regulations for radiation exposure through drinking water, stating that current standards are impractical to enforce. The EPA is recommending that intervention is not necessary until drinking water is contaminated with radioactive iodine 131 at a concentration of 81,000 picocuries per liter (the limit for short term exposure set by the International Atomic Energy Agency), which is 27,000 times the current EPA limit of 3 picocuries per liter for long term exposure.

National Pollutant Discharge Elimination System

The National Pollutant Discharge Elimination System (NPDES) permit program addresses water pollution by regulating point sources that discharge pollutants to waters of the United States. Created in 1972 by the Clean Water Act, the NPDES permit program is authorized to state governments by EPA to perform many permitting, administrative, and enforcement aspects of the program.

Radiation Protection

EPA has the following seven project groups to protect the public from radiation.

1. Radioactive Waste Management
2. Emergency Preparedness and Response Programs

   Protective Action Guides And Planning Guidance for Radiological Incidents

   EPA developed the manual to provide guideline for local and state governments to protect public from nuclear accident.

1. EPA Cleanup and Multi-Agency Programs
2. Risk Assessment and Federal Guidance Programs
3. Naturally-Occurring Radioactive Materials Program
4. Air and Water Programs
5. Radiation Source Reduction and Management

Tools for Schools

The rate of asthma in children is increasing at a pace that underscores the need for schools to address indoor air problems. EPA's Indoor Air Quality Tools for Schools Program helps maintain a healthy environment and reduce exposures to indoor environmental contaminants. The Program helps school personnel identify, solve, and prevent indoor air quality problems in the school environment. Through the use of a multi-step management plan and checklists for the entire building, schools can also lower their students' and staff's risk of exposure to asthma triggers.

Environmental Education

The National Environmental Education Act of 1990 requires EPA to provide national leadership to increase environmental literacy. EPA established the Office of Environmental Education to implement this program.

Environmental Impact Statement Reviews

EPA is responsible for reviewing Environmental Impact Statements of other federal agencies' projects, under the National Environmental Policy Act (NEPA).

Clean School Bus USA

Clean School Bus USA is a national partnership to reduce children's exposure to diesel exhaust by eliminating unnecessary school bus idling, installing effective emission control systems on newer buses and replacing the oldest buses in the fleet with newer ones. Its goal is to reduce both children's exposure to diesel exhaust and the amount of air pollution created by diesel school buses.

Research Vessel

OSV *Bold* docked at Port Canaveral, FL

On March 3, 2004, the United States Navy transferred USNS *Bold*, a *Stalwart* class ocean surveillance ship, to the EPA, now known as OSV *Bold*. The ship, previously used in anti-submarine operations during the Cold War, is equipped with sidescan sonar, underwater video, water and sediment sampling instruments, used in study of ocean and coastline. One of the major missions of the *Bold* was to monitor for ecological impact sites where materials are dumped from dredging operations in U.S. ports. In 2013, the *Bold* was awarded to Seattle Central Community College (SCCC) by the General Services Administration. SCCC demonstrated in a competition that they would put it to the highest and best purpose, and acquired the ship at a cost of $5,000.

Advance Identification

Advance identification, or ADID, is a planning process used by the EPA to identify wetlands and other bodies of water and their respective suitability for the discharge of dredged and fill material. The EPA conducts the process in cooperation with the U.S. Army Corps of Engineers and local states or Native American Tribes. As of February 1993, 38 ADID projects had been completed and 33 were ongoing.

Freedom of Information Act Processing Performance

In the latest Center for Effective Government analysis of 15 federal agencies which receive the most Freedom of Information Act FOIA requests, published in 2015 (using 2012 and 2013 data, the most recent years available), the EPA earned a D by scoring 67 out of a possible 100 points, i.e. did not earn a satisfactory overall grade.

Controversies

EPA headquarters in Washington, D.C.

There has been political controversy over whether environmental regulations generally increase or decrease national employment.

The "LT2" Drinking Water Controversy

In 2005, the EPA issued the Long Term 2 Enhanced Surface Water Treatment Rule (LT2). The rule requires covering open-air reservoirs containing finished drinking water, in order to reduce the incidence of disease caused by microorganisms in drinking water.

To comply with the rule the EPA ordered that a cap be placed over the Hillview Reservoir, the open reservoir where water bound for New York City's receives its final disinfection before entering the pipelines that serve the city. A number of city and state officials complained that the project was too costly and unnecessary. The Hillview project is currently being reviewed by the EPA.

In Portland, Oregon, there was opposition to capping its reservoirs from a local citizens group and city officials, but recently the city has decided to stop fighting and bring its reservoirs into compliance with LT2.

Fiscal Mismanagement

EPA director Anne M. Gorsuch resigned under fire in 1983 during a scandal over mismanagement of a $1.6 billion program to clean up hazardous waste dumps. Gorsuch based her administration of the EPA on the New Federalism approach of downsizing federal agencies by delegating their functions and services to the individual states. She believed that the EPA was over-regulating business and that the agency was too large and not cost-effective. During her 22 months as agency head, she cut the budget of the EPA by 22%, reduced the number of cases filed against polluters, relaxed Clean Air Act regulations, and facilitated the spraying of restricted-use pesticides. She cut the total number of agency employees, and hired staff from the industries they were supposed to be regulating. Environmentalists contended that her policies were designed to placate polluters, and accused her of trying to dismantle the Agency.

In 1982 Congress charged that the EPA had mishandled the $1.6 billion toxic waste Superfund and demanded records from Gorsuch. Gorsuch refused and became the first agency director in U.S. history to be cited for contempt of Congress. The EPA turned the documents over to Congress several months later, after the White House abandoned its court claim that the documents could not be subpoened by Congress because they were covered by executive privilege. At that point, Gorsuch resigned her post, citing pressures caused by the media and the congressional investigation. Critics charged that the EPA was in a shambles at that time.

Fuel Economy

In July 2005, an EPA report showing that auto companies were using loopholes to produce less fuel-efficient cars was delayed. The report was supposed to be released the day before a controversial energy bill was passed and would have provided backup for those opposed to it, but at the last minute the EPA delayed its release.

The state of California sued the EPA for its refusal to allow California and 16 other states to raise fuel economy standards for new cars. EPA administrator Stephen L. Johnson claimed that the EPA was working on its own standards, but the move has been widely considered an attempt to shield the auto industry from environmental regulation by setting lower standards at the federal level, which would then preempt state laws. California governor Arnold Schwarzenegger, along with governors from 13 other states, stated that the EPA's actions ignored federal law, and that *existing* California standards (adopted by many states in addition to California) were almost twice as effective as the *proposed* federal standards. It was reported that Stephen Johnson ignored his own staff in making this decision.

After the federal government bailed out General Motors and Chrysler in the Automotive industry crisis of 2008–2010, the 2010 Chevrolet Equinox was released with EPA fuel economy rating abnormally higher than its competitors. Independent road tests found that the vehicle did not out-perform its competitors, which had much lower fuel economy ratings. Later road tests found better, but inconclusive, results.

Palm-based biodiesel and renewable diesel failed to meet the minimum 20% greenhouse gas (GHG) emissions savings threshold requirement to qualify as renewable fuels under the US Renewable Fuel Standard 2. Palm oil plantations threaten the habitats of the endangered orangutang and pygmy elephant.

Global Warming

In December 2007, EPA Administrator Stephen L. Johnson approved a draft of a document that declared that climate change imperiled the public welfare—a decision that would trigger the first national mandatory global-warming regulations. Associate Deputy Administrator Jason Burnett e-mailed the draft to the White House. White House aides—who had long resisted mandatory regulations as a way to address climate change—knew the gist of what Johnson's finding would be, Burnett said. They also knew that once they opened the attachment, it would become a public record, making it controversial and difficult to rescind. So they did not open it; rather, they called Johnson and asked him to take back the draft. Johnson rescinded the draft; in July 2008, he issued a new version which did not state that global warming was danger to public welfare. Burnett resigned in protest.

Libraries

In 2004, the Agency began a strategic planning exercise to develop plans for a more virtual approach to library services. The effort was curtailed in July 2005 when the Agency proposed a $2.5 million cut in its 2007 budget for libraries. Based on the proposed 2007 budget, the EPA posted a notice to the Federal Register, September 20, 2006 that EPA Headquarters Library would close its doors to walk-in patrons and visitors on October 1, 2006. The EPA also closed some of its regional libraries and reduced hours in others, using the same FY 2007 proposed budget numbers.

On October 1, 2008, the Agency re-opened regional libraries in Chicago, Dallas and Kansas City and the library at its Headquarters in Washington, DC.

In June 2011, the EPA Library Network published a strategic plan for fiscal years 2012–2014.

Mercury Emissions

In March 2005, nine states (California, New York, New Jersey, New Hampshire, Massachusetts, Maine, Connecticut, New Mexico and Vermont) sued the EPA. The EPA's inspector general had determined that the EPA's regulation of mercury emissions did not follow the Clean Air Act, and that the regulations were influenced by top political appointees. The EPA had suppressed a study it commissioned by Harvard University which contradicted its position on mercury controls. The suit alleges that the EPA's rule allowing exemption from "maximum available control technology" was illegal, and additionally charged that the EPA's system of pollution credit trading allows power plants to forego reducing mercury emissions. Several states also began to enact their own mercury emission regulations. Illinois's proposed rule would have reduced mercury emissions from power plants by an average of 90% by 2009.

9/11 Air Ratings

An August 2003 report released by EPA's Inspector General claimed that the White House put pressure on the EPA to delete cautionary information about the air quality in New York City around Ground Zero following the September 11, 2001 attacks.

An Environmental Protection Agency employee checks one of the many air sampling locations set up around the World Trade Center site.

Very Fine Airborne Particulates

Tiny particles, under 2.5 micrometres, are attributed to health and mortality concerns, so some health advocates want the EPA to regulate it. The science may be in its infancy, although many conferences have discussed the trails of this airborne matter in the air. Foreign governments such as Australia and most Member state of the European Union have addressed this issue.

The EPA first established standards in 1997, and strengthened them in 2006. As with other standards, regulation and enforcement of the $PM_{2.5}$ standards is the responsibility of the state governments, through State Implementation Plans.

Political Pressure and Scientific Integrity

In April 2008, the Union of Concerned Scientists said that more than half of the nearly 1,600 EPA staff scientists who responded online to a detailed questionnaire reported they had experienced incidents of political interference in their work. The survey included chemists, toxicologists, engineers, geologists and experts in other fields of science. About 40% of the scientists reported that the interference had been more prevalent in the last five years than in previous years. The highest number of complaints came from scientists who were involved in determining the risks of cancer by chemicals used in food and other aspects of everyday life.

EPA research has also been suppressed by career managers. Supervisors at EPA's National Center for Environmental Assessment required several paragraphs to be deleted from a peer-reviewed journal article about EPA's integrated risk information system, which led two co-authors to have their names removed from the publication, and the corresponding author, Ching-Hung Hsu, to leave EPA "because of the draconian restrictions placed on publishing". EPA subjects employees who author scientific papers to prior restraint, even if those papers are written on personal time. A $3 million mapping study on *sea level rise* was suppressed by EPA management during both the Bush and Obama Administrations, and managers changed a key interagency report to reflect the removal of the maps. EPA employees have reported difficulty in conducting and reporting the results of studies on hydraulic fracturing due to industry and governmental pressure, and are concerned about the censorship of environmental reports.

In 2015, the Government Accountability Office stated that the EPA violated federal law with covert propaganda on their social media platforms. The social media messaging that was used promoted materials supporting the Waters of the United States rule, including materials that were designed to oppose legislative efforts to limit or block the rule.

Environmental Justice

The EPA has been criticized for its lack of progress towards environmental justice. Administrator Christine Todd Whitman was criticized for her changes to President Bill Clinton's Executive Order 12898 during 2001, removing the requirements for government agencies to take the poor and minority populations into special consideration when making changes to environmental legislation, and therefore defeating the spirit of the Executive Order. In a March 2004 report, the inspector general of the agency concluded that the EPA "has not developed a clear vision or a comprehensive strategic plan, and has not established values, goals, expectations, and performance measurements" for environmental justice in its daily operations. Another report in September 2006 found the agency still had failed to review the success of its programs, policies and activities towards environmental justice. Studies have also found that poor and minority populations were underserved by the EPA's Superfund program, and that this situation was worsening.

Barriers to Enforcing Environmental Justice

Localization

Many issues of environmental justice are localized, and are therefore hard to be addressed by federal agencies such as the EPA. Without significant media attention, political interest, or 'crisis' status, local issues are less likely to be addressed at the local or federal level. With a still developing sector of environmental justice under the EPA, small, local incidents are unlikely to be solved compared to larger, well publicized incidents.

Conflicting Political Powers

The White House maintains direct control over the EPA, and its enforcements are subject to the political agenda of who is in power. Republicans and Democrats differ in their approaches to, and perceived concerns of, environmental justice. While President Bill Clinton signed the executive order 12898, the Bush administration did not develop a clear plan or establish goals for integrating environmental justice into everyday practices, which in turn affected the motivation for environmental enforcement.

Responsibilities of the EPA

The EPA is responsible for preventing and detecting environmental crimes, informing the public of environmental enforcement, and setting and monitoring standards of air pollution, water pollution, hazardous wastes and chemicals. While the EPA aids in preventing and identifying hazardous situations, it is hard to construct a specific mission statement given its wide range of responsibilities. It is impossible to address every environmental crime adequately or efficiently if there is no specific mission statement to refer to. The EPA answers to various groups, competes for resources, and confronts a wide array of harms to the environment. All of these present challenges, including a lack of resources, its self-policing policy, and a broadly defined legislation that creates too much discretion for EPA officers.

Authority of the EPA

Under different circumstances, the EPA faces many limitations to enforcing environmental justice. It does not have the authority or resources to address injustices without an increase in federal mandates requiring private industries to consider the environmental ramifications of their activities.

Gold King Mine Waste Water Spill

On August 5, 2015, while examining the level of pollutants in the Gold King Mine, EPA workers released over three million gallons of toxic waste water, including heavy metals such as lead and arsenic into Cement Creek, which flowed into the Animas River in Colorado.

Local Emergency Planning Committee

Local Emergency Planning Committees (LEPCs) are community-based organizations that assist in preparing for emergencies, particularly those concerning hazardous materials. Under the Emergency Planning and Community Right-to-Know Act (EPCRA), Local Emergency Planning Committees (LEPCs) must develop an emergency response plan, review the plan at least annually, and provide information about hazardous materials in the community to citizens. Plans are developed by LEPCs with stakeholder participation. The LEPC membership must include (at a minimum):

- Elected state and local officials
- Police, fire, civil defense, and public health professionals
- Environment, transportation, and hospital officials
- Facility representatives
- Representatives from community groups and the media

Some required elements of the community emergency response plan, developed by the LEPC, include:

- Identification of facilities and transportation routes of extremely hazardous substances
- Description of emergency response procedures, on and off site

- Designation of a community coordinator and facility emergency coordinator(s) to implement the plan
- Outline of emergency notification procedures
- Description of how to determine the probable affected area and population by releases
- Description of local emergency equipment and facilities and the persons responsible for them
- Outline of evacuation plans
- A training program for emergency responders (including schedules)
- Methods and schedules for exercising emergency response plans

Though LEPCs were created with the Federal law EPCRA, through the U.S. Environmental Protection Agency, they are often funded partially by the U.S. Department of Transportation's Hazardous Materials Emergency Preparedness grant program. Other sources of funding may include local jurisdictions, industry, businesses, NGOs, and other public or private grants.

World Conference on Disaster Risk Reduction

conferences focusing on disaster and climate risk management in the context of sustainable development. The World Conference has been convened three times, with each edition to date having been hosted by Japan: in Yokohama in 1994, in Kobe in 2005 and in Sendai in 2015. As requested by the UN General Assembly, the United Nations Office for Disaster Risk Reduction (UNISDR) served as the coordinating body for the Second and Third UN World Conference on Disaster Reduction in 2005 and 2015.

The conferences bring together government officials and other stakeholders, such as NGOs, civil society organizations, local government and private sector representatives from around the world to discuss how to strengthen the sustainability of development by managing disaster and climate risks. The Third UN World conference adopted the Sendai Framework for Disaster Risk Reduction 2015-2030. Previous conference outcomes include the Hyogo Framework for Action 2005 - 2015: Building the Resilience of Nations and Communities to Disasters in 2005 and the Yokohama Strategy and Plan of Action for a Safer World in 1994.

Contents

- 3 2015 Third UN World Conference on Disaster Risk Reduction (WCDRR) in Sendai
    - o 3.1 Sendai Framework for Disaster Risk Reduction 2015-2030

1994 First World Conference on Natural Disasters in Yokohama

The First World Conference on Natural Disasters in Yokohama, Japan from 23 to 27 May 1994, adopted the Yokohama Strategy for a Safer World: Guidelines for Natural Disaster Prevention, Preparedness and Mitigation and its Plan of Action, endorsed by the UN General Assembly in 1994. It was the main outcome of the mid-term review of the International Decade of Natural Disaster Reduction (IDNDR) and established 10 principles for its strategy, a plan of action and a follow-up. Furthermore, it provides guidelines for natural disaster prevention, preparedness and mitigation.

Ten principles of the Yokohama Strategy for a Safer World

The ten principles of the Yokohama Strategy for a Safer World:

1. Risk assessment is a required step for the adoption of adequate and successful disaster reduction policies and measures.
2. Disaster prevention and preparedness are of primary importance in reducing the need for disaster relief.
3. Disaster prevention and preparedness should be considered integral aspects of development policy and planning at national, regional, bilateral, multilateral and international levels.
4. The development and strengthening of capacities to prevent, reduce and mitigate disasters is a top priority area to be addressed during the Decade so as to provide a strong basis for follow-up activities to the Decade.
5. Early warnings of impending disasters and their effective dissemination using telecommunications, including broadcast services, are key factors to successful disaster prevention and preparedness.
6. Preventive measures are most effective when they involve participation at all levels, from the local community through the national government to the regional and international level.
7. Vulnerability can be reduced by the application of proper design and patterns of development focused on target groups, by appropriate education and training of the whole community.
8. The international community accepts the need to share the necessary technology to prevent, reduce and mitigate disaster; this should be made freely available and in a timely manner as an integral part of technical cooperation.
9. Environmental protection as a component of sustainable development consistent with poverty alleviation is imperative in the prevention and mitigation of natural disasters.

10. Each country bears the primary responsibility for protecting its people, infrastructure, and other national assets from the impact of natural disasters. The international community should demonstrate strong political determination required to mobilize adequate and make efficient use of existing resources, including financial, scientific and technological means, in the field of natural disaster reduction, bearing in mind the needs of the developing countries, particularly the least developed countries.

2005 Second World Conference on Disaster Reduction in Kobe

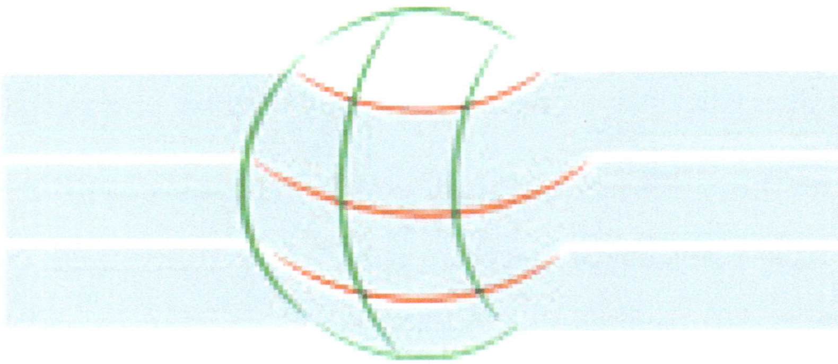

The Second World Conference on Disaster Reduction conference was held in Kobe, Japan from 18 to 22 January 2005. This conference took on particular poignancy, as it came almost 10 years to the day after the Great Hanshin earthquake in Kobe and less than a month after the 2004 Indian Ocean tsunami. Japan's long history of severe natural disasters, prominence in international humanitarian aid and development and its scientific achievements in monitoring dangerous natural phenomena also made it a suitable conference venue.

The upcoming conference had not garnered much attention, but due to the December 26 Indian Ocean tsunami, the attendance grew dramatically and the international media focused on the event. Japan's Emperor Akihito opened the conference and welcomed 4,000 participants from around the world.

The World Conference adopted plans to put in place an International Early Warning Programme (IEWP), which had first been proposed at the Second International Conference on Early Warning in 2003 in Bonn, Germany.

Conference Objectives

The goal of the World Conference was to find ways to reduce the toll of disasters through preparation, and ultimately to reduce human casualties. Due to the proximity to the devastating Indian Ocean tsunami, developing a global tsunami warning system was high on the agenda. Other topics included:

- pledges to reduce disaster damage
- healthcare after disaster
- early warning systems

- safe building standards
- agree upon cost-effective preventative countermeasures
- a global database on relief and reconstruction and a centre on water hazards

The Pacific Rim Tsunami Warning system is an example of a cost-effective warning system; its yearly operating cost is approximately US$4 million. The yearly operating cost of a hypothetical global warning system is estimated at US$30 million. This cost, compared to the international aid donations of nearly US$8 billion for the 2004 Indian Ocean earthquake and tsunami, clearly demonstrates the cost effectiveness of such a system.

Hyogo Framework for Action

The Hyogo Framework for Action (2005–2015): Building the Resilience of Nations and Communities to Disasters was an outcome of the 2005 conference. The Hyogo Framework (HFA) was the first plan to explain, describe and detail the work required from all different sectors and actors to reduce disaster losses. It was developed and agreed on with the many partners needed to reduce disaster risk - governments, international agencies, disaster experts and many others - bringing them into a common system of coordination. The HFA, which ran from 2005 to 2015, set five specific priorities for action:

1. Making disaster risk reduction a priority;
2. Improving risk information and early warning;
3. Building a culture of safety and resilience;
4. Reducing the risks in key sectors;
5. Strengthening preparedness for response.

2015 Third UN World Conference on Disaster Risk Reduction (WCDRR) in Sendai

UN World Conference on
Disaster Risk Reduction
2015 Sendai Japan

The Third UN United Nations World Conference on Disaster Risk Reduction was held in Sendai, Japan from 14 to 18 March 2015, drawing 6,500 delegates to the conference itself and 50,000 people to the associated Public Forum. Sendai is the largest city of Miyagi Prefecture, in north-eastern Japan. It has a prominent status as it was hit by the Great East Japan earthquake, 130 kilometres from the epicentre. The conference included discussion of the aftermath of the Japanese response to the 2011 disaster and how Japan's early warning system can save lives when earthquakes and tsunamis strike. The conference included an announcement of a US$4 billion fund to prepare for disasters over four years. Moreover, the conference coincided with Cyclone Pam hitting Vanuatu, and Vanuatu President Baldwin Lonsdale urgently requested international assistance for his people. Thus, in 2015, many developed nations announced they would partner with smaller countries to prepare for future disaster relief operations.

The conference adopted the Sendai Framework for Disaster Risk Reduction 2015-2030. The Sendai Framework is the first major agreement of the post-2015 development agenda, with seven targets and four priorities for action. It was endorsed by the UN General Assembly in June 2015.

Sendai Framework for Disaster Risk Reduction 2015-2030

The Sendai Framework is a 15-year non-binding agreement which recognizes that the State has the primary role to reduce disaster risk but that responsibility should be shared with other stakeholders including local government and the private sector. It aims for the following outcome:

"The substantial reduction of disaster risk and losses in lives, livelihoods and health and in the economic, physical, social, cultural and environmental assets of persons, businesses, communities and countries."

The Sendai Framework emerged from three years' of consultations and negotiations, supported and coordinated by UNISDR, during which UN member states, NGOs and other stakeholders made calls for an improved version of the existing Hyogo Framework, with a set of common standards, a comprehensive framework with achievable targets, and a legally-based instrument for disaster risk reduction. Based on the Sendai Framework, the first was held in New Delhi as part of Asian ministerial conference for disaster risk reduction. Member states also emphasised the need to tackle disaster risk reduction and climate change adaption when setting the Sustainable Development Goals, particularly in light of an insufficient focus on risk reduction and resilience in the original Millennium Development Goals.

The Sendai Framework sets four specific priorities for action:

1. Understanding disaster risk;
2. Strengthening disaster risk governance to manage disaster risk;
3. Investing in disaster risk reduction for resilience;
4. Enhancing disaster preparedness for effective response, and to "Build Back Better" in recovery, rehabilitation and reconstruction.

To support the assessment of global progress in achieving the outcome and goal of the Sendai Framework, seven global targets have been agreed:

1. Substantially reduce global disaster mortality by 2030, aiming to lower average per 100,000 global mortality between 2020-2030 compared to 2005-2015;

2. Substantially reduce the number of affected people globally by 2030, aiming to lower the average global figure per 100,000 between 2020-2030 compared to 2005-2015;
3. Reduce direct disaster economic loss in relation to global gross domestic product by 2030;
4. Substantially reduce disaster damage to critical infrastructure and disruption of basic services, among them health and educational facilities, including through developing their resilience by 2030;
5. Substantially increase the number of countries with national and local disaster risk reduction strategies by 2020;
6. Substantially enhance international cooperation to developing countries through adequate and sustainable support to complement their national actions for implementation of the framework by 2030;
7. Substantially increase the availability of and access to multi-hazard early warning systems and disaster risk information and assessments to the people by 2030.

Federal Disaster Assistance Nonprofit Fairness Act of 2013

The Federal Disaster Assistance Nonprofit Fairness Act of 2013 (H.R. 592) is a bill that passed in the United States House of Representatives during the 113th United States Congress. The bill would make religious organizations and religious non-profits eligible to receive federal funding for repairs and rebuilding of their facilities after a major disaster. The bill amends the Robert T. Stafford Disaster Relief and Emergency Assistance Act.

The bill passed the House by a large margin, but was criticized by opponents for using taxpayer money to help tax-exempt organizations and for violating the principle of separation of church and state.

Provisions/Elements of the Bill

The Bill would amend the Robert T. Stafford Disaster Relief and Emergency Assistance Act to include community centers, including tax-exempt houses of worship, as "private nonprofit facilities" for purposes of disaster relief and emergency assistance eligibility under such Act.

The amendment makes a church, synagogue, mosque, temple, or other house of worship, and a private nonprofit facility operated by a religious organization, eligible for federal contributions for the repair, restoration, and replacement of facilities damaged or destroyed by a major disaster, without regard to the religious character of the facility or the primary religious use of the facility.

Finally, the bill makes the Act applicable to the provision of assistance in response to a major disaster or emergency declared on or after October 28, 2012.

Procedural History

House

The Bill was introduced in the House of Representatives on February 8, 2013 by Rep. Chris Smith (R-NJ). The bill was immediately referred to the United States House Committee on Transportation and Infrastructure. It was then referred to the United States House Transportation Subcommittee on Economic Development, Public Buildings and Emergency Management on February 11, 2013. On February 13, the Bill was considered on the House floor under a suspension of the rules; it passed later that day 354-72 (Roll Call Vote 39). By the time it had passed, the Bill had received 9 cosponsors in the House. Of the 72 members who voted against the bill, 66 were Democrats and 6 were Republicans.

Senate

The Bill was received in the Senate on February 14, 2013. It was referred to the United States Senate Committee on Homeland Security and Governmental Affairs on March 13, 2013.

Debate

Supporters of the Bill argued that the bill was necessary to clarify existing law by correcting an oversight. The House Republican Majority's official website stated their view that "unlike previous disasters, Congress did not clarify that religious nonprofits, including houses of worship, are eligible to receive disaster funding when providing aid for victims and communities impacted by Hurricane Sandy. As a result, the (Obama) Administration is interpreting current law to preclude houses of worship from being eligible for funding on the same terms as similarly impacted nonprofits." Other supporters of federal funding for houses of worship damaged in disasters have argued that such places a key parts of communities and are not exempted from other aspects of government disaster management, like evacuation orders or emergency fortification attempts, and therefore should not be prevented from seeking disaster relief.

Opponents of the Bill argued that the Obama Administration is right to deny funds to religious institutions; this argument was often made on First Amendment grounds. The non-profit Center for Inquiry wrote on its website that they opposed the Bill because they believed it violated the First Amendment's prohibition against government establishing any religion. The Center for Inquiry argued that "true religious freedom protects the conscience of the taxpayer by ensuring that his or her money is not used to support or advance religion with which he or she may disagree." One blogger referred to the act as the "Taxpayer Money to Tax-Exempt Churches Act". The group Americans United for Separation of Church and State also argued against the Bill, writing on their website that "Church-state separation protects the right of taxpayers to support only the religious institutions of their choice, and it ensures the independence and integrity of faith communities. Houses of worship exist primarily to teach the tenets of their religion. They should be supported by donations, not government subsidies."

References

- Turner, D.B. (1994). Workbook of atmospheric dispersion estimates: an introduction to dispersion modeling (2nd ed.). CRC Press. ISBN 1-56670-023-X.

- Making the Grade: Access to Information Scorecard 2015 March 2015, 80 pages, Center for Effective Government, retrieved 21 March 2016
- Delhi (2016-11-03). "Prime Ministers address at Asian Ministerial Conference on Disaster Risk Reduction". Business Standard India. Retrieved 2016-11-27.
- Milman, Oliver (2016-02-26). "Republican candidates' calls to scrap EPA met with skepticism by experts". The Guardian. ISSN 0261-3077. Retrieved 2016-11-22.
- Adler, Jonathan H. (2015-12-15). "GAO hits EPA for 'covert propaganda' to promote 'waters of the United States' (WOTUS) rule". The Washington Post. ISSN 0190-8286. Retrieved 2015-12-16.
- Schlanger, Zoë (August 7, 2015). "EPA Causes Massive Spill of Mining Waste Water in Colorado, Turns Animas River Bright Orange". Retrieved August 10, 2015.
- Kolb, Joseph J. (August 10, 2015). "'They're not going to get away with this': Anger mounts at EPA over mining spill". Retrieved August 10, 2015.
- "Yokohama Strategy and Plan of Action, Guidelines for Natural Disaster Prevention, Preparedness and Mitigation" (PDF). UNISDR. Retrieved 27 August 2015.
- "World Conference on Disaster Reduction opens in Kobe, Press Release". UN New Centre. 18 Jan 2005. Retrieved 27 August 2015.
- "Highlights for Tuesday, 18 January 2005 on the Second World Conference for Disaster Reduction 2005". IISD. 18 Jan 2005. Retrieved 27 August 2015.
- "Global Early Warning System launched at Conference on Disaster Reduction, Press Release". UN New Centre. 19 January 2005. Retrieved 27 August 2015.
- "Five years after the tsunami in the Indian Ocean" (PDF). United Nations Educational, Scientific and Cultural Organization, Intergovernmental Oceanographic Commission. 2009. Retrieved 27 August 2015.
- "Hyogo Framework for Action 2005-2015: Building the Resilience of Nations and Communities to Disasters" (PDF). UNISDR. 22 Jan 2005. Retrieved 27 August 2015.
- "Ban praises Sendai earthquake recovery as model for turning tragedy of disaster into sustainable future". UN News Centre. 15 Mar 2015.
- Elizabeth Chuck (March 15, 2015). "'Disaster' in Vanuatu After Cyclone Pam Tears Through Pacific Archipelago". NBC News. Retrieved March 15, 2015.
- "Japan Launches US$4 Billion Sendai Cooperation Initiative As Third UN World Conference On Disaster Risk Reduction Opens". Relief Web. 14 Mar 2015.

Permissions

All chapters in this book are published with permission under the Creative Commons Attribution Share Alike License or equivalent. Every chapter published in this book has been scrutinized by our experts. Their significance has been extensively debated. The topics covered herein carry significant information for a comprehensive understanding. They may even be implemented as practical applications or may be referred to as a beginning point for further studies.

We would like to thank the editorial team for lending their expertise to make the book truly unique. They have played a crucial role in the development of this book. Without their invaluable contributions this book wouldn't have been possible. They have made vital efforts to compile up to date information on the varied aspects of this subject to make this book a valuable addition to the collection of many professionals and students.

This book was conceptualized with the vision of imparting up-to-date and integrated information in this field. To ensure the same, a matchless editorial board was set up. Every individual on the board went through rigorous rounds of assessment to prove their worth. After which they invested a large part of their time researching and compiling the most relevant data for our readers.

The editorial board has been involved in producing this book since its inception. They have spent rigorous hours researching and exploring the diverse topics which have resulted in the successful publishing of this book. They have passed on their knowledge of decades through this book. To expedite this challenging task, the publisher supported the team at every step. A small team of assistant editors was also appointed to further simplify the editing procedure and attain best results for the readers.

Apart from the editorial board, the designing team has also invested a significant amount of their time in understanding the subject and creating the most relevant covers. They scrutinized every image to scout for the most suitable representation of the subject and create an appropriate cover for the book.

The publishing team has been an ardent support to the editorial, designing and production team. Their endless efforts to recruit the best for this project, has resulted in the accomplishment of this book. They are a veteran in the field of academics and their pool of knowledge is as vast as their experience in printing. Their expertise and guidance has proved useful at every step. Their uncompromising quality standards have made this book an exceptional effort. Their encouragement from time to time has been an inspiration for everyone.

The publisher and the editorial board hope that this book will prove to be a valuable piece of knowledge for students, practitioners and scholars across the globe.

Index